FOREWORD

Throughout the OECD Member countries, many national programmes on radioactive waste management are considering geological disposal in argillaceous media. In order to determine their suitability for waste disposal, it is necessary to evaluate of the potential migration of radionuclides from such a disposal system to the accessible environment. Those evaluations require not only the site-specific data of a site-programme, but also a sound general understanding of the basic physical and chemical processes that govern solute transport through those formations.

In that context, the NEA Working Group on Measurement and Physical Understanding of Groundwater Flow Through Argillaceous Media (informally called the "Clay Club") has been established to address the many issues associated with that subject. The Working Group promotes a constant intercomparison of the properties of the different argillaceous media under consideration for geological disposal, as well as an exchange of technical and scientific information by means of meetings, workshops and written overviews on relevant subjects.

The definition of the chemical and isotopic composition of groundwater present in argillaceous formations considered as potential host rock for radioactive waste disposal is crucial notably for:

i) establishing their properties as a barrier to radionuclide migration;

ii) understanding the disturbances that may be induced during the excavation of the repository facility and by the presence of the waste and the engineered-barrier system;

iii) defining the type of water that will interact with the engineered-barrier system.

The lack of standard protocols and certified standards on how to perform porewater extractions from argillaceous formations has led the "Clay Club" to launch a critical review of the relevant literature on the current methods applied to extract water and solutes and on the various available approaches to interpret their results. That desk study was commissioned to the *Laboratoire d'Hydrologie et de Géochimie Isotopique* (UMR OrsayTerre, CNRS–Université de Paris-Sud, France) by a consortium of national organisations represented within the "Clay Club".

This document provides a synthesis of available porewater extraction methods, assesses their respective advantages and limitations, identifies key processes that may influence the composition of the extracted water, describes modelling approaches used to determine in situ porewater composition, and highlights, wherever possible, some unresolved issues and recommendations on ways to address them.

The opinions and conclusions expressed are those of the authors only, and do not necessarily reflect the views of the funding organisations, any OECD Member country or international organisation. This report is published on the responsibility of the Secretary General of the OECD.

Radioactive Waste Management

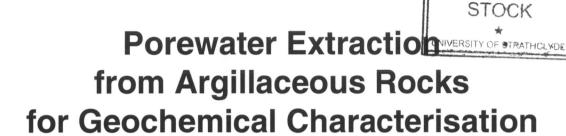
Porewater Extraction
from Argillaceous Rocks
for Geochemical Characterisation

Methods and Interpretation

NUCLEAR ENERGY AGENCY
ORGANISATION FOR ECONOMIC CO-OPERATION AND DEVELOPMENT

ORGANISATION FOR ECONOMIC CO-OPERATION AND DEVELOPMENT

Pursuant to Article 1 of the Convention signed in Paris on 14th December 1960, and which came into force on 30th September 1961, the Organisation for Economic Co-operation and Development (OECD) shall promote policies designed:

–   to achieve the highest sustainable economic growth and employment and a rising standard of living in Member countries, while maintaining financial stability, and thus to contribute to the development of the world economy;
–   to contribute to sound economic expansion in Member as well as non-member countries in the process of economic development; and
–   to contribute to the expansion of world trade on a multilateral, non-discriminatory basis in accordance with international obligations.

The original Member countries of the OECD are Austria, Belgium, Canada, Denmark, France, Germany, Greece, Iceland, Ireland, Italy, Luxembourg, the Netherlands, Norway, Portugal, Spain, Sweden, Switzerland, Turkey, the United Kingdom and the United States. The following countries became Members subsequently through accession at the dates indicated hereafter: Japan (28th April 1964), Finland (28th January 1969), Australia (7th June 1971), New Zealand (29th May 1973), Mexico (18th May 1994), the Czech Republic (21st December 1995), Hungary (7th May 1996), Poland (22nd November 1996) and the Republic of Korea (12th December 1996). The Commission of the European Communities takes part in the work of the OECD (Article 13 of the OECD Convention).

NUCLEAR ENERGY AGENCY

The OECD Nuclear Energy Agency (NEA) was established on 1st February 1958 under the name of the OEEC European Nuclear Energy Agency. It received its present designation on 20th April 1972, when Japan became its first non-European full Member. NEA membership today consists of 27 OECD Member countries: Australia, Austria, Belgium, Canada, Czech Republic, Denmark, Finland, France, Germany, Greece, Hungary, Iceland, Ireland, Italy, Japan, Luxembourg, Mexico, the Netherlands, Norway, Portugal, Republic of Korea, Spain, Sweden, Switzerland, Turkey, the United Kingdom and the United States. The Commission of the European Communities also takes part in the work of the Agency.

The mission of the NEA is:

–   to assist its Member countries in maintaining and further developing, through international co-operation, the scientific, technological and legal bases required for a safe, environmentally friendly and economical use of nuclear energy for peaceful purposes, as well as
–   to provide authoritative assessments and to forge common understandings on key issues, as input to government decisions on nuclear energy policy and to broader OECD policy analyses in areas such as energy and sustainable development.

Specific areas of competence of the NEA include safety and regulation of nuclear activities, radioactive waste management, radiological protection, nuclear science, economic and technical analyses of the nuclear fuel cycle, nuclear law and liability, and public information. The NEA Data Bank provides nuclear data and computer program services for participating countries.

In these and related tasks, the NEA works in close collaboration with the International Atomic Energy Agency in Vienna, with which it has a Co-operation Agreement, as well as with other international organisations in the nuclear field.

ACKNOWLEDGEMENTS

The national organisations represented within the NEA Working Group on Measurement and Physical Understanding of Groundwater Flow Through Argillaceous Media ("Clay Club") and the NEA wish to express their gratitude to the authors of this report, E. Sacchi and J.-L. Michelot from the *Laboratoire d'Hydrologie et de Géochimie Isotopique* (Université de Paris-Sud, France). They also wish to commend the special contribution of H. Pitsch from the French Atomic Energy Commission (CEA), whose laboratory hosted E. Sacchi for the duration of the project.

This document has been jointly supported by a consortium of national organisations represented within the "Clay Club":

- ANDRA, France (National Radioactive Waste Management Agency);

- CEA, France (Atomic Energy Commission);

- CEN/SCK, Belgium (Nuclear Energy Research Centre).

- ENRESA, Spain (Spanish National Agency for Radioactive Waste);

- GRS, Germany (Company for Reactor Safety)

- IPSN, France (Atomic Energy Commission/Nuclear Protection and Safety Institute);

- NAGRA, Switzerland (National Co-operative for the Disposal of Radioactive Waste); and

- ONDRAF/NIRAS, Belgium (Belgian Organisation for Radioactive Waste and Fissile Materials).

All those organisations are sincerely thanked for their support and for their valuable reviews and comments.

We also wish to acknowledge for their contribution and helpful discussion: A.M. Fernández (CIEMAT, Spain), M. Hagwood (Schlumberger, The Hague) and M. Dusseault (PMRI, Canada).

J.F. Aranyossy (ANDRA, France) has been the promoter, within the "Clay Club", of this study project.

P. Lalieux has been in charge of the co-ordination of this report on behalf of the NEA (Radiation Protection and Waste Management Division).

TABLE OF CONTENTS

EXECUTIVE SUMMARY

This Executive Summary is a synopsis of the report *"Extraction of Water and Solutes from Argillaceous Rocks for Geochemical Characterisation: Methods and Critical Evaluation"* commissioned by the OECD/Nuclear Energy Agency Working Group on Measurement and Physical Understanding of Groundwater Flow Through Argillaceous Media ("Clay Club") to the *Laboratoire d'Hydrologie et de Géochimie Isotopique* (UMR OrsayTerre CNRS-Université de Paris-Sud, France). Yet, it may also be considered as an independent document accessible to the non-specialised, technical reader. The references quoted in the Executive Summary are the most representative of the issue discussed in the text.

INTRODUCTION

The definition of the chemical and isotopic composition of the groundwater present in argillaceous formations that are considered as potential host rock for radioactive waste disposal, is crucial notably for establishing their properties as a barrier to radionuclide migration, for understanding the disturbances that may be induced during the excavation of the repository and later by the presence of the waste and the engineered barrier system, and for defining the type of water that will interact with the latter.

Historically, water and solute extraction techniques for geochemical characterisation have been developed for petroleum geology, pedology and unsaturated and saturated zone hydrology. Their application to fine-grained sediments, especially if rich in clay minerals and low in water content, is delicate. That has been shown by comparative studies (see for example Walker *et al.*, 1994) using different techniques on the same sample or the same technique on different types of sample. In fact, chemical and isotope fractionation is very often observed and those effects are poorly reproducible. Thus, fundamental questions arise, such as: what is the degree of representativity of the measured composition? Does it account for the whole porewater or only part of it? Furthermore, standard protocols and certified norms on how to perform porewater extractions from argillaceous formations are missing. Hence, the decision by the "Clay Club" to launch a critical review of the relevant literature on currently applied porewater extraction methods and on the various approaches to the interpretation of their results.

The critical review report provides a synthesis of available porewater extraction methods, assesses their respective advantages and limitations, identifies key processes that may influence the composition[1] of the extracted water, describes modelling approaches that are used to determine *in-situ* porewater composition, and highlights, wherever possible, some unresolved issues and recommendations on ways to address them. The report is intended to cover work published or submitted

---

1.   In this Executive Summary and the review report, the term "composition" refers to both the chemical and isotopic composition of the porewater sample, except if otherwise mentioned.

for publication before January 1999, but conclusions also take into account results from internal reports and unpublished work known to the authors.

The first part of the report is a review of the fundamental clay properties that are considered as responsible for the non-linear response of the clay/water/solute system during the extraction of the solution. The second part is based on an exhaustive bibliographical study of the available extraction techniques, with a focus on applications to clay-rich media. For each method, description and examples of applications are presented, in order to determine the advantages and problems of each technique as a function of the purpose of the investigation. The aim of the third part of the report is to analyse the mechanisms involved in water and solute extraction processes and the possible consequences on the isotopic and chemical composition of the extracted clay porewater. An indirect derivation approach, extensively based on geochemical modelling, is presented at the end of that chapter. Finally, in the conclusions, indications are given for further experiments, both for the definition of clay/water/solute interactions and for intercomparative studies.

The report covers the whole range of argillaceous media currently considered for deep disposal, *i.e.,* from soft, potentially plastic clays with relatively high water content, to hard, potentially fractured mudrocks with low to very low water content. Targeted applications are the marls of the Palfris formation (Wellenberg, Switzerland), the Opalinus clay (Switzerland), the Boom clay (Mol, Belgium), the Jurassic mudrocks of the Paris Basin (France) and the Toarcian argillites crossed by the Tournemire tunnel (France).

A.   THE CLAY/WATER SYSTEM

The word "clay" has been generally used for indicating fine-grained sediments (particle size less than 2 µm) with particular attributes of plasticity (Guggenheim and Martin, 1995). Clays are mainly constituted by a specific group of aluminosilicates, consisting in alternations of tetrahedral and octahedral sheet structures, held together by electrostatic forces. Most of the properties displayed by clay minerals are due to their small grain size and their sheet-like shape, both factors causing a very high surface area compared to the mass of material (Van Olphen, 1963).

A.1   Interactions between water, solutes and clay

The water molecule owes most of its solvent efficiency to its polarity. In order to minimise the energy of the dipole, the water molecule tends to rotate and create bonds with other ions, molecules and mineral surfaces. The number of water molecules that may be attracted by an ion in the solution and the strength of the established bonds vary according to the ratio of the cation radius and its charge. The stability of the hydration complex may be measured by the exchange rate and/or the mobility of the water molecule in the hydration shell of the ion. Some of them, especially those of high-charge, low-radius ions are to be considered rather stable (*e.g.,* magnesium).

Clay particles, like all colloids, tend to develop electric charges when in contact with a liquid phase. In addition, a deficit of positive charge caused by isomorphic substitutions inside the clay lattice is frequently found. As a consequence, positive ions and water molecules are attracted towards the clay surface. The arrangement of the water molecules in the neighbouring region may be disturbed by the electric field, and differ from that of bulk water. Thus, the thermodynamic, hydrodynamic and spectroscopic properties of water in clays may differ from those of pure water (Sun *et al.,* 1986).

The effect of cation hydration is very important for the structure and properties of hydrated clays: cations with a low hydration energy, such as K^+, NH_4^+, Rb^+ and Cs^+ produce interlayer dehydration and layer collapse, and are therefore fixed in the interlayer positions. Conversely, cations with high hydration energy such as Ca^{2+}, Mg^{2+} and Sr^{2+} produce expanded layers and are more easily exchanged (Sawhney, 1972).

Clay hydration involves adsorption of a number of water molecules on the exposed surfaces of clay particles. Three adsorption processes seem to take place with the increase in water content, corresponding to different types of water:

(a) adsorption of water in the interlayer space, inside clay particles (Type I), called "interlamellar water";

(b) continuous hydration related to unlimited adsorption of water around primary clay particles (Type II), called "intraparticle water";

(c) free-water condensation in micropores (Type III), called "interparticle water".

The first hydration stage, corresponding to very low water contents, occurs on the hydrophilic sites of the clay surface. Kaolinite group minerals develop low surface charges, and therefore only adsorb up to a monolayer of water molecules in the interlayers. At higher water contents, the monolayer of water molecules is formed, and bonds may be established between the water molecules. Given the constraints of the silicate structure, bonds have to adjust by stretching and rotating, consequently slowing the motion of water molecules. In smectites, the presence of a cation adsorbed on the surface to balance electric-charge deficiencies is extremely important (Anderson and Low, 1958). In fact, it appears that the first hydration stage proceeds through cation hydration.

Capillary condensation at the contact points between particles and/or grains constitutes the main adsorption mechanism in the previously defined second domain, giving rise to Type-II water in the interparticle pores. That water is still bound to the clay particles and forms layers of water molecules whose thickness is proportional to the water activity. As the total water content increases, free-water condenses in the pores (Type-III water).

Another fundamental property of clays is the cation selectivity (Thomas, 1977). When a solution and clay come into contact and reach equilibrium, some ions are adsorbed preferably to others. In general, divalent ions are preferred in swelling-clay interlayers over monovalent ions, while the opposite may be true for some high-charge clays. Cation partitioning depends on many parameters, such as the charge unbalance with respect to the charge of the ion, the interlayer spacing or the dimension of the adsorbing site (surface functional group) with respect to the ionic radius, the composition of the solution and temperature. As clay particles normally develop a negative charge, anions will tend to be repulsed. That phenomenon is known as anion exclusion. Anion retention, especially for phosphate, fluorine and sulphate ions, may occur on particular reactive sites. In addition, natural organic matter shows a strong affinity for clay minerals (Theng, 1974; Maurice et al., 1998).

In summary, the position, size and charge of interlamellar exchangeable cations largely determine the spatial arrangement of the water molecules. Recent studies on the water mobility in clay-rich systems support the notion that the water is more influenced by the saturating cation than by the particular clay (Weiss and Gerasimowicz, 1996).

A.2   Porosity in clays

Studies on the size distribution and accessibility of pores to solutions showed that the degree of water "immobilisation" is greater near the pore walls than in the centre of the pore. If pores are less than a few tens of nanometres in diameter, the water and the solutes cannot flow unless a threshold pressure gradient is exceeded. In addition, because of the negative charges developed by the clay mineral surfaces, some regions of the pores may be precluded access to negative ions.

As a consequence, beside the physical porosity, that is the ratio of the void volume to the total volume, other types of porosity may be defined (Pearson, in press). In particular, the porosity of the water volume effectively available for the migration of each ion or molecule is called "geochemical porosity", and represents the fluid volume in which reactions occur. It is required for geochemical and reactive transport modelling. It is similar to transport porosities and, in clay-rich materials, is close to diffusion porosity.

In coarse-grained rocks, all porosity types are approximately equal because of the lack or the minor influence of attractive and repulsive forces exerted by the solid phase. Pearson (1998) reports estimates of those different porosity types for clay-rich rocks (London clay, clay-rich Canadian tills, Boom clay, Opalinus clay and Palfris marl). According to those calculations, geochemical and diffusion porosities for water molecules are the same and are equal to water-content porosity. Geochemical and diffusion porosities for solutes that do not have access to interlayer or surface sorbed waters constitute only one-third to one-half of the water content porosity. Those porosity values are shown to vary with the salinity of the solution (Karnland, 1997).

A.3   Organic matter

Various types of organic substances may be found in sediments, including aliphatic and aromatic hydrocarbons and non-hydrocarbons geopolymers known as bitumen and kerogen (Yariv and Cross, 1979). The organic matter characterisation of rocks may provide information on the biological input, the palaeodepositional environment, and the degree of maturity and degradation. Three fractions may be distinguished within the organic matter: the soluble fraction (dissolved organic carbon, DOC), that may be found in porewaters, the solvent extractable fraction (bitumen) and the non-extractable fraction (kerogen). Those three fractions show different properties and relate in different ways to the main topic of this study.

Humic substances, representing most of the soluble organic matter, display a variety of functional groups with different reactivities. They may develop an anionic character, tend to form strong hydrogen bonds with water molecules and may associate intermolecularly, changing molecular conformation in response to changes in pH, redox conditions, electrolyte concentration and functional group binding. The degree of complexity resulting from those properties is much larger for humic substances than for other biomolecules, as they reflect the behaviour of interacting polymeric molecules instead of the behaviour of a structurally well-defined single type of molecule.

The soluble organic matter is known to form strong complexes with metal ions (Yong et al., 1992). Organic matter bound to clay particles presents a reactive surface to solutes. As a consequence, it may play a role in the buffering of proton and metal cation concentration in the porewater solution via cation exchange. The association between the solid phases and the insoluble organic matter is not very well understood because of the poor definition of the organic matter structure. Nevertheless, the same functional groups responsible for cation exchange are also thought to be responsible for the binding of the organic matter to the clay particle. A review of the clay/organic matter interactions may be found in Theng (1974).

A.4 Conclusions

The studied system is constituted by clay minerals, water molecules, dissolved ions and molecules, and the organic matter. All those components interact with each other, creating bonds of different energies. When trying to extract the porewater solution, the applied energy will need to break those bonds. The extraction conditions will, in most cases, not be uniform throughout the sample, but will depend on the local desaturation level. While the water content decreases, new bonds between the components may be created, and other strengthened. An additional problem is the reduced porosity of the system, acting as a filter and thus accelerating or retarding the ions according to their radius and charge. All those effects, combined together, are responsible for the non-linear response of the system, and for the chemical and isotopic fractionation effects observed during the extraction procedures. Therefore, a quantitative description of those effects remains a complex and, in most cases, unresolved problem.

B. EXPERIMENTAL METHODS

B.1 Field techniques for fluid extraction and characterisation

B.1.1 Piezometer and borehole drilling

In-situ water-extraction techniques normally involve drilling. Drilling operations should be carefully planned in order to avoid long-lasting contamination of the environment with drilling fluids (Tshibangu *et al.*, 1996). In fact, in very low water-content systems, long delays are required before purging the system from that type of contamination, and cases have been reported where a representative formation fluid could never be recovered (NAGRA, 1997).

In order to reduce and limit contamination, a few suggestions to consider include:

–   Using an "equilibrated" drilling fluid to which a tracer may be added to follow the extent and duration of the contamination;

–   Using air drilling instead of conventional drilling, especially for short dedicated boreholes;

–   Avoiding, whenever possible, the use of casing and backfill material. If necessary, consider PVC and stainless-steel casings. In soft clays, self-sealing piezometers may be installed, exploiting the natural convergence of the clay.

–   Enhancing the collection of the solution in the piezometers by sealing it with a packer and slightly underpressurising the borehole. The types of piezometers installed at Mont Terri are good examples of non-contaminating water-sampling devices.

In-situ physico-chemical measurements (pH, Eh and T)

Ion-selective electrodes have been extensively studied for the determination of the fluid composition in high water-content systems (see Frant, 1997 for a review). Recent developments of microelectrodes (De Wit, 1995; Kappes *et al.*, 1997) have not yet found widespread application in low water-content systems. They are rather used for establishing "*in-situ*" profiles in sediments (Hales *et al.*,

1994; Hales and Emerson, 1996) or for measuring pH and other parameters on very small quantities of extracted water (*e.g.,* on water extracted from squeezing cells).

Among the numerous physico-chemical parameters to be considered, pH is of particular interest for the chemistry regulation of interstitial fluids. Early attempts of *in-situ* measurements using glass electrodes directly in the piezometers failed because of pressure problems (glass electrodes are very fragile) and calibration problems (Griffault *et al.,* 1996). Electrodes installed outside the borehole with a circulation pump homogenising the fluid may be easily isolated from the circuit for calibration, without inducing major disturbances of the system (see *e.g.,* Mont Terri project). pH measurement within the borehole may also be obtained using a fibre-optic pH sensor (Motellier *et al.,* 1995). That system was tested against other pH measurement techniques (batch and in-flow with a glass electrode) at the HADES facility (Pitsch *et al.,* 1995b) and showed a very good agreement with them. In addition to its pressure resistance, the device has a good response even if the water around the sensing tip is not renewed. Consequently, it may also be used in situations where discharge is lower than 1 ml/h (Pitsch *et al.,* 1995b).

Attempts to obtain an *in-situ* Eh measure have been made, but results seem to be mainly restricted to fluids that display a measurable equilibrium potential imposed by a dissolved redox couple (buffered solutions) or a stable mixed potential imposed by two different redox couples. In the absence of electroactive species, the measured potential is unstable because very small disturbances (concentration fluctuations, electrode surface modifications) cause the potential to change abruptly from one value to the other (Pitsch *et al.,* 1995a). A flow cell technique tested on samples collected in different clay/water environments proved to be more reliable than batch measurements, even for reducing environments, provided that anaerobic conditions are respected. More reliable Eh values may be obtained by modelling and/or extrapolating the data obtained for the redox couples that are known to be present in the fluid, even if not in sufficient quantity to provide a stable electrode potential ($< \sim 5$ μM) (Beaucaire *et al.,* 1998).

The use of an optical fibre for borehole-temperature logging is reported by Förster *et al.* (1997). Although resolution is 5 to 10 times lower than conventional techniques, that system quickly responds and is not affected by problems related to variations in cable resistance, that disturb electric currents, and to improper isolation.

Field techniques for indirect fluid characterisation

Borehole-logging tools represent a mature technology that is widely used in oil and gas exploration (Schlumberger, 1997). The application of those techniques to radioactive-waste repositories assessment is currently being investigated. A combination of wireline logging, seismic, hydrologic and geomechanical testing techniques may provide valuable information for site characterisation. That includes fracture and fault detection and mapping, the physical properties of the rock (lithology, stratigraphy, porosity), geochemistry (rock-forming elements), hydrologic properties (conductivity, transmissivity), *in-situ* stress and geomechanical properties.

Since the 1960s, Schlumberger and other service companies (Kenyon *et al.,* 1995) have applied nuclear magnetic resonance (NMR) to the *in-situ* determination of rock porosity, moisture content and amount of free and bound water. The *in-situ* NMR technique has also been applied successfully to the oil industry and seems to give very promising results, even for fine-grained sediments. That technique is currently undergoing testing on low-permeability clay formations at the ONDRAF and NAGRA sites (Win *et al.,* 1998; Strobel *et al.,* 1998). Uncertainties are related to the estimation of constants necessary to calculate permeability; those constants are based on semi-empirical test results established for sandstones. An accurate calibration of field NMR data to core permeabilities, together with grain size analysis, may improve the tool ability to log those parameters on a continuous basis. Finally, the integration of that tool with the other logging tools should successfully quantify hydrologic properties and detect heterogeneities in argillaceous formations. It may be considered as an alternative or complement to additional coring (Croussard *et al.*, 1998).

B.2 Rock sampling, storage and preservation

The Eh and pH conditions existing in any deep rock formation have a significant influence on almost all critical parameters determining the system behaviour with respect to radionuclide migration. The process of sampling, that isolates portions of rock or fluid from their environment, may induce changes in the sample characteristics that may be virtually impossible to estimate with any certainty. Closely associated with that problem are the issues of long-term storage, handling and preparation of samples.

In low water-content systems, artefacts seem to proceed very slowly, due to the low diffusion coefficients of the species in the small porosity. Effects such as dehydration and oxidation may be prevented with some simple precautions: isolating as soon as possible the sample from the atmosphere, performing all the pre-treatment operations (crushing, sieving, etc.), by minimising the contact time with the atmosphere and the use of potentially contaminating tools or devices. Sample may be conditioned by wrapping the rock in aluminium foil and beeswax, or coating it directly with paraffin. Aluminium-plastic bags or foils that may be flushed with nitrogen or mixtures of inert gases, then evacuated and thermally welded may also be used. That type of conditioning seems adequate to preserve samples for mineralogical and chemical analyses, but does not entirely protect the solution from the risk of evaporation. Wrapping the rock in aluminium is highly recommended if the organic matter has to be analysed.

It is recommended, in order to eliminate the possibility of a contamination from the drilling fluid, that the outside rim of the cores is trimmed and discarded.

B.3 On-sample laboratory techniques

The most important techniques considered in the study are:

- Centrifugation: the sample is placed in a closed container, then spun in a centrifuge at a given number of rotations per minute (rpm). The pressure difference developed across the sample exceeding the capillary tension holding the water in the pores causes the water to be extracted (Batley and Giles, 1979). Additionally, heavy liquids immiscible with the solution, may be used. They percolate through the pores, pushing out the solution found floating on the top of the sample.

- Squeezing: in most cases, the sample is crushed, then placed in a hydraulic press and squeezed. The pore fluid is expelled through a stainless-steel filter and collected outside the press with a syringe.

- Leaching: the crushed sample is placed in contact with deionised water or another solution, at a given solid/liquid ratio. After establishing equilibrium, the solid phase is separated and the liquid phase is analysed. The resulting composition is interpreted with different modelling levels (from simple mixing between deionised water and porewater to more complex models).

- Vacuum distillation: the crushed rock sample is placed in a preparation line that is evacuated and heated. The water is extracted by evaporation and the released water molecules are collected in frozen traps.

- Azeotropic distillation: that process is based on the observation that some solvents (toluene, xylene, petroleum ether, etc.) form an azeotropic mixture with water, featuring a boiling point lower than the boiling points of the two end members. The crushed sample is placed in a flask, immersed in the selected solvent and gradually heated. At the boiling point of the azeotrope, the mixture evaporates, recondensing in the funnel with a cloudy appearance. At room temperature, the two liquids (water and solvent) separate.

- Direct equilibration: those techniques are based on the equilibrium established between the porewater in the rock sample and a known amount of a given substance with a known isotopic composition. It does not involve the physical extraction of the porewater, but simply an equilibration through the gaseous phase.

As different extracting principles are used, not all the techniques may provide solutions for both chemical and isotopic analyses. In the following table (Table A), the techniques are listed, together with indications (based on bibliography) on their suitability for chemical and isotopic analyses.

Table A. **Use of different water extraction techniques for chemical and isotopic analyses**

Technique	Specifications	Chemical analysis	Isotopic analysis
Centrifugation	Low/high speed	Major and trace elements	^{18}O and ^{2}H
	Heavy liquids	Major and trace elements	Not investigated
Squeezing		Major and trace elements	^{18}O and ^{2}H
Leaching		Major and trace elements	^{2}H and ^{3}H
Distillation	Under vacuum	Impossible	^{18}O and ^{2}H
	Azeotropic	Impossible	^{18}O and ^{2}H
Direct equilibration	With CO_2	Impossible	^{18}O only
	With water	Impossible	^{18}O and ^{2}H

Each technique is extensively discussed in the text, with particular reference to the technical specifications, the percentage of recovered water, application examples and artefacts mentioned in the literature. All those data are summarised in Table B. In addition, the maximum applied suction, measuring the bond-breaking strength and expressed in pF^2, is reported. As a reference, it should be remembered that, beyond 11 layers of adsorbed water molecules, the properties of water resemble those

---

2.   pF is the decimal logarithm of the suction expressed in centimetres (head) of water.

Table B. **List of water-extraction techniques and their known artefacts**

Technique	Specifications	Maximum. applied suction	Recognised artefacts	Advantages and possible applications
Centrifugation	Low speed (2,500 rpm)	pF 3	Solution oxidation	Suitable for high water content sediments
	High speed (14,000 rpm)	pF 4.8	Solution oxidation; decrease in concentration of the extracted solution with increasing extraction	
	Ultracentrifugation (20,000 rpm) + solvent displacement	pF 4.4	Danger of organic-matter destruction	Better controlling redox sensitive elements
Squeezing	Low pressure (5 MPa)	pF 4.7	Small or undetectable	Suitable for high water-content sediments and clays
	High pressure (70 MPa)	pF < 5.8	According to different authors, small or undetectable up to 60-100 MPa; with increasing pressure, both concentration increases and decreases of solutes are reported	To be validated
	High pressure (552 MPa)	pF < 6.7		Not completely validated
Leaching	Deionised water	–	Dissolution of minerals; cation exchange with the clay	Possibly good for obtaining chloride from highly saline porewaters
	High selectivity complexes	–	Complete exchange with the adsorbed cations	If coupled with modelling, may provide cation occupancies on the clay
Distillation	Vacuum distillation	pF 7	Possibly incomplete extraction; non-reliable ^{18}O values	If the extraction is complete, deuterium may provide useful information as a water-molecule tracer
	Azeotropic distillation	pF 7	Possibly incomplete extractions; systematic depletion of deuterium values	Not completely validated
Direct equilibration	With CO_2	–	Possibly incomplete equilibration, as equilibration times are difficult to estimate	^{18}O only. Apparently reliable
	With water	–		Promising technique, needs validation

of bulk water (Swartzen-Allen and Matijevic, 1974; Sposito and Prost, 1982). In addition, according to Van Olphen (1965), five layers of water on montmorillonite display a suction potential corresponding to a pF of 4.7, two layers to a pF of 6.4, and one layer to a pF of 6.7. The air-inlet point, corresponding to the beginning of desaturation, for a clay such as Boom clay is situated at a pF of approximately 4 (Horseman *et al.*, 1996). Logically, that value should be higher for indurated clays with lower water content.

According to the calculated suction values, apart from low pressure squeezing and centrifugation, all the techniques should be able to extract free water from clays, and most of the techniques are likely to affect also any water strongly bound to the clay surfaces. The possibility of extracting not only free water, but also to some extent strongly bound water, is a crucial issue. In fact, the impossibility to extract all the solution or to control the relative amounts prevents the derivation of a "true" porewater composition.

Useful information on the amount of different types of water, together with the suction parameters to be considered for extracting only the free water, may be derived from various studies on adsorption-desorption isotherms (Decarreau, 1990), nuclear magnetic resonance (NMR), infrared spectroscopy (IRS) (Prost, 1975; Sposito and Prost, 1982) and dielectric relaxation. That type of investigation should be routinely performed in order to characterise the clay/water environment.

C.   PROCESSES AND CURRENT INTERPRETATIONS

The disturbance-inducing processes related to water and solute extractions are reviewed in this section of the report, together with the recognised artefacts and the attempts made to correct them. All those effects, although they concern in theory all the extraction techniques, are more relevant to some of them. In Table C, all the data are reported, together with an estimate of their influence on the extracted solution and the possibility to correct them. The table shows clearly that there is a risk of obtaining a non-representative sample in most of the techniques. As a consequence, there is little doubt that many arguments may be raised against all the investigations conducted with those methods and the results obtained.

The basic problem relates obviously to the presence of different types of water in the clay/water system. From the hydrogeological point of view, only the free water (in amount and composition) is of interest, because it represents the fraction possibly mobilised under given hydraulic conditions. Nevertheless, in the type of rocks considered in this review, with low water content, the amount of free water is probably so small that each attempt to extract it may involve having to deal with the strongly bound water as well. That may be due to the dishomogeneity of the water distribution inside the sample, affecting the local conditions of water availability, as well as to the slow movement, due to diffusion of the water molecule itself.

Ideally, the approaches adopted so far have either claimed to extract only the requested type of water, or tried to extract all the water and calculate the porewater composition, assuming the behaviour of the water/rock system. Unfortunately, both those options are not verified, and in each study case we have evidence of partial extraction of different solution types, the relative amounts of which are unknown.

What information may we consider as reliable then? In our opinion, very few. Those are mainly distribution profiles across the studied clay formations of chloride obtained by leaching, deuterium obtained by distillation (provided that no serious salinity differences are detected in the

Table C. List of processes occurring during water extraction and possible corrections of their effects on water composition

Physical process	Techniques	Major effects	Possible corrections
Increase in pressure	Squeezing, centrifugation	Decrease in concentration of the extracted solution with increasing pressure; modified solubility of the solid phases; possible effects also on the isotopic composition of the solution.	At present, no satisfactory model allows the correction of the data. More detailed experiments (ongoing) are needed to validate the technique.
Decrease in pressure	Mainly related to sampling and conditioning	Degassing of the solution, possibly leading to the precipitation of carbonates and the reduction of porosity.	Thermodynamic modelling of the carbonate system allows an estimation of the artefacts.
Oxidation	Related to bad sampling and conditioning; also induced by centrifugation, leaching, and to a minor extent by squeezing if the operations are not conducted in a controlled atmosphere.	Oxidation of solid phases, mainly sulphides; change in pH of the solution; dissolution of carbonate minerals; modified stability of other phases; cation exchange with clays and organic matter.	Impossible to correct because of the complexity of the system. Data should be discarded.
Change in temperature	All techniques. Most of the squeezing and centrifugation devices are currently equipped with temperature controls.	No major consequences if the temperature differences are low (< 10°C); modified stability of the solid phases if greater.	Apparently, most of the effects are reversible with storage at the original temperature.
Ion exchange	Mainly leaching, but possibly induced by all techniques.	Major changes in the solution composition; dissolution and precipitation of solid phases.	May be evaluated and possibly corrected via geochemical modelling.
Salt dissolution	Potentially all techniques increasing pressure and water to rock ratio, especially leaching.	Major changes in the solution composition; cation exchange.	May be evaluated if the mineralogy is well known.
Salt precipitation	Potentially all techniques decreasing water to rock ratio.	Major changes in the solution composition; cation exchange; possible modification of the isotopic composition if hydrated phases are precipitated.	May be estimated by geochemical modelling and by the mineralogical observation of the dry sample.
Incomplete water extraction	All techniques except leaching	Non-representativity of the chemical and isotopic composition of the solution.	Difficult to estimate without a deeper understanding of the clay/water system. May be modelled and corrected for stable isotopes.

porewaters) or equilibration, and to some extent, noble gas measurements. An extensive discussion is conducted in the text to justify those statements.

C.1   Geochemical modelling

An interesting new approach to obtain information on the porewater composition has been developed and is extensively based on thermodynamic modelling. The continuous development and refinement of the thermodynamic data bases, coupled with the increasing performances of computer codes, allows at present the modelling of most of the dissolution/precipitation reactions and part of the cation-exchange reactions.

Two experimental approaches are currently adopted in the framework of the investigations in clay-rich environments. One, suggested by Bradbury et al. (1990) is an experimental procedure extensively based on rock-sample leaching with different solutions (deionised water, high affinity complexes) and on the thermodynamic processing of the data. On the basis of the considerations made on all the ions in the aqueous and the high-affinity complex solution extracts, the quantity of highly soluble salts (sodium chloride, potassium chloride) is derived, together with the cation-exchange capacity of the rock and the ion occupancy on the exchange sites. Subsequently, the calculation of the cation ratios in the liquid phase, that are in equilibrium with known occupancies, is performed, knowing the selectivity coefficients for the different cations. That methodological approach is completely different as it encompasses both the technical problem of water extraction and the definition of the amount and characteristics of the "free" and "bound" solution. Since its design, it has been applied on samples of the Palfris marl (Baeyens and Bradbury, 1991; 1994), and is currently used in the framework of the investigations for the Mont Terri project (Bradbury and Baeyens, 1997; Bradbury et al., 1997; Bradbury and Baeyens, 1998; Pearson et al., in preparation). Although uncertainties remain because the clay/water system is not univocally defined, reasonable assumptions on porewater composition may be made and checked with in-situ equilibration.

Another modelling approach that has been proven valid for clay environments has been proposed by Beaucaire et al. (1995; in press) on the Boom clay. Their approach relies on the regional groundwater characterisation (Boom-clay porewater and groundwater from the Rupelian aquifer). The acquisition and regulation mechanisms of the groundwater and porewater compositions at the regional scale are considered to be the same, and the fluids to have a common origin. By a careful observation of the correlation between major cations and anions, a mixing process between porewater and a marine solution may be identified. However, a simple mixing model between those two end members does not describe the system precisely, suggesting that exchange and equilibration with the host rock also occur. Beaucaire et al. used a dissolution-precipitation model, considering a mineral assemblage of solid phases that have all been identified in the Boom clay and whose dissolution equilibria are well established, in order to predict the water composition. The model seems to describe accurately the variability of the recognised types of water within the regional scale. Discrepancies between predicted and measured concentrations of major elements are within analytical uncertainties except for very dilute species; as for pH, the deviation is less than 0.3 unit. That good predictivity induced De Windt et al. (1998b) to test the model on the Tournemire water collected from a draining fracture of the site and resulted in a quite satisfactory agreement between modelled and analysed water compositions.

Ion-exchange and dissolution/precipitation processes are both known to occur during water/rock interaction, but at different timescales: ion exchanges are quickly established, while equilibrium is more slowly attained for dissolution/precipitation reactions. Of course, both aspects need to be considered in the safety assessment procedure for waste-disposal sites. This means that a considerable work of thermodynamic-data generation is to be carried out in order to reach definite conclusions and to elaborate models that take into account simultaneously both aspects of the water/rock interaction.

C.2   Conclusions, recommendations and topics for further investigation

The problem of extracting solutions from argillaceous formations for geochemical and isotopic characterisations is complex, as expected. For the time being, the presence of different forces arising from the clay/water interaction and influencing the movement of water molecules and solutes prevents the possibility to define experimentally the "true" porewater composition. That composition is needed for several objectives:

- To perform corrosion calculations for canisters and matrices where radioactive waste will be contained.

- To evaluate the age and the natural movement of water and solutes across the formation.

- To calculate the speciation and the solubility of phases in order to evaluate the water/rock interaction phenomena affecting radionuclide migration.

- To foresee the effect of the site water on the engineered barriers of the repository.

Fortunately, not all those objectives require the same degree of knowledge of the porewater composition.

For corrosion studies, the main parameters needed are the total salinity, the oxidising or reducing properties of the solution and the speciation of particular elements (*e.g.,* sulphur). In the host rock, it may be reasonably assumed that, given the low solid/water ratio, the solution composition would soon be controlled by the minerals in the material of interest, the composition and reactivity of which is fairly well known.

For tracing the water age and movement, a few techniques have proven reliable. Noble gases measurements may provide information on groundwater age. Deuterium and chloride, provided that they are interpreted in relative, rather than in absolute, terms and that porosity characteristics are well known, allow for an estimation of the time required to establish the distribution profile, as long as the movement is diffusion-dominated. Tritium may be used for tracer experiments, as apparently it may be easily extracted by vacuum distillation and measured accurately by radioactive counting.

Speciation studies are the most affected by the analytical problems we have encountered. Here, the whole task of characterising the water content and composition relies on the absence of a clear definition of which part of the cations and anions belongs to the clay surface (being adsorbed and strongly bound) and which part belongs to the bulk solution. So far, that problem has been neglected or treated in terms of total water content. Without that definition, the question would turn up as to which total suction needs to be applied to our sample in order to extract even more water, that is actually no porewater at all. As a consequence, future investigations should aim at a better understanding of the fundamental properties of the clay/water system: thermodynamics of pore-confined water is still a critical issue.

Table D summarises those concluding remarks.

Table D. **Examples of reliable information according to the investigation**

Critical issues of radioactive waste management	Information provided by	Reliable analytical techniques	Limitations
Interaction with the barriers	Solid phases + thermodynamic data	Leaching + geochemical modelling	Potentially not so predictable for short term effects
Age and groundwater movement	Noble gases	Diffusion in evacuated containers	None
	Deuterium distribution profiles	Vacuum distillation Equilibration	Careful check of all parameters: interpretation in relative terms
	Chloride distribution profiles + isotopes	Leaching	Potentially not suitable for low salinity systems: interpretation in relative terms
Speciation and radionuclide migration	Solution composition + rock properties	Geochemical modelling/ + leaching	Check with *in-situ* equilibration

More fundamental research is needed to understand the physico-chemical processes involved during water extraction. Information may be obtained by:

–   Coupling the mechanical behaviour with the mineralogical and chemical characteristics of the system. Macroscopic properties, such as swelling and mechanical strength, depend on the water content but also on the type of saturating cation on the clay surface and consequently on the porewater composition.

–   Conducting rigorous physical studies on the pore size and distribution, using relatively "soft" techniques, that reduce to the extent possible pore-size modifications during the study.

–   Evaluating the amount of free and bound water. That should be achieved through indirect techniques such as NMR, IRS and dielectric relaxation spectroscopy (DRS). Those techniques have the advantage to cause fewer disturbances to the clay sample. In addition, recent technical developments allow the use of those techniques directly inside exploratory boreholes, bypassing all the artefacts related to sample collection and preservation.

–   Validating some new and promising techniques through the design of experiments on a variety of clay environments and testing their applicability to different mixtures of clay minerals and different salinities of the interstitial solutions.

Other specific topics where further investigation is needed concern the behaviour of chlorine during the extraction by leaching, the rock characterisation by its ion-exchange isotherms and the production of reliable exchange constants for thermodynamic modelling, as well as the definition of the water-movement mechanisms within the sample during the extraction by squeezing. Furthermore, a great amount of work needs to be done in the field of organic-matter extraction, characterisation and evaluation of its retention role.

Considering isotopic studies, an improvement of existing techniques and/or the development of new techniques, such as direct equilibration, seem possible. A validation of the squeezing technique for isotopic analysis of the extracted solutions is also necessary.

However, more essentially, the critical issue in water extraction for geochemical interpretation is to relate accurately the chemical and isotopic composition of the extracted water to the *in-situ* water. Due to the complexity of the processes involved in each method, the extracted water is always an image, and only an image, of the porewater. It is therefore a challenge for the physicochemist to explain the fluid transformations during extraction, in order to back-calculate the *in-situ* water from its image and provide the geochemist with the needed tools.

An intercomparison of analytical techniques on clays has already been organised by the NEA (Van Olphen and Fripiat, 1979) and a second one was launched by the IAEA on isotope analysis (Walker *et al.*, 1994). In agreement with what is previously stated, there is little doubt in our minds that, if those benchmark experiments were conducted again today, the results would not be anything else but the same. At the present stage, intercomparisons must only be considered as one of the many tools for understanding better the phenomenology of each analytical method. No conclusion on reliability may be drawn from any intercomparison exercise as long as no fundamental knowledge may assess the accuracy of one or another method.

PREFACE

SCOPE, OBJECTIVES AND LIMITS OF THE STUDY

The need to obtain information on the porewater composition of sedimentary rocks, especially those targeted for waste isolation purposes, has been clear since the beginning of the investigations on that type of formations. Firstly, the distribution of dissolved constituents in the formation results from the geologic history of the massif, the transport processes affecting the fluids and the water/rock interaction. The discrimination between those different processes, when possible, enables to establish the boundary conditions of the system. Secondly, the definition of a groundwater in chemical equilibrium with the host rock, is a prerequisite for beginning any credible sorption study (Baeyens and Bradbury, 1991). Besides, that composition will most probably constitute an end-member of any mixing with fluids (shallow waters, drilling fluids, etc.) that will most likely occur during the excavation of a repository. Consequently, porewater composition is needed in all studies concerning the disturbance of the deep environment before the waste isolation and the calculations of the time required to restore the initial conditions. Finally, that will represent the type of water that will be in contact with the engineered-barrier system (*e.g.,* concrete lining, backfill materials, metallic waste packages, glass) and, after interaction with the latter, it may be responsible for the leaching and transport of the radionuclides present in the waste.

Historically, water and solute extraction techniques from sedimentary rocks for chemical and isotopic analyses have been developed for petroleum geology, pedology and unsaturated and saturated zone hydrology. Their application to fine-grained sediments, especially if rich in clay minerals and low in water content, is delicate. That has been shown by comparative studies (see for example Walker *et al.*, 1994) using different techniques on the same sample or the same technique on different sample types. In fact, chemical and isotope fractionation is very often observed when the soil or rock is rich in clay minerals and very poor in interstitial water. Besides, those fractionation effects are poorly reproducible. Thus, the question of sample representativity arises: different types of solutions are present in the clayey rock, corresponding to different types of bonding between water molecules, dissolved ions and clay particles. The issue of sample representativity is crucial in the framework of the assessment of the performances of a repository located in argillaceous host rocks as isotopic and geochemical data, for example, are used to support and test flow models in those geological media.

Numerous papers on the topic have been published, but attempts of synthesis are very rare and mainly concern applications to soils.

This report aims to be a comprehensive critical review of the extraction techniques of water and solutes from argillaceous rocks, for chemical and isotopic analyses, and the available approaches to interpret their results. The study is subdivided in three main parts.

Following a first part reviewing the fundamental of clay properties, the second part of the report is based on an exhaustive bibliographical study of the available extraction techniques, with a focus on applications to clay-rich media. Both *in-situ* techniques (from piezometers, boreholes or special underpressurised equipment) and laboratory techniques on sample (centrifugation, squeezing, leaching, distillation, etc.) are considered. For each water extraction method, description and examples of applications are presented. Chemical and isotopic data obtained in each case are examined in order to determine the advantages and problems of each technique in relation to the investigation.

The third part of the report aims to analyse the mechanisms involved in water and solutes extraction processes, as well as the possible consequences on the isotopic and chemical composition of the extracted clay porewater. Finally, short indications are given for further experiments, both for the definition of clay/water/solute interactions and for intercomparative studies.

This document attempts to set a basis for an international methodological effort on results obtained from different investigation sites. It covers the whole range of argillaceous media currently considered for deep disposal, *i.e.,* from soft, potentially plastic clays with relatively high water content, to hard, potentially fractured mudrocks with low to very low water content. Targeted applications are the Palfris formation (Wellenberg, Switzerland), the Opalinus clay (Switzerland) the Boom clay (Mol, Belgium), the Jurassic mudrocks of the Paris Basin (France) and the Toarcian formation crossed by the Tournemire Tunnel (France).

The work has been conducted in close link with the French Atomic Energy Commission (CEA/LIRE) for the critical interpretation of chemical results. Discussion with other scientists commonly using those techniques within the framework of their investigations allowed for an overview of the problems and difficulties most commonly encountered. It also helped to focus the objectives of the study, identify the areas where more investigations are needed and widen the perspective to a more general international concern. Contacts were also taken with the organisations represented within the "Clay Club" and with practical activities in that field. That helped assessing the current state of the art. In an early phase, those organisations have been requested to provide their input to the list of bibliographic references serving as a basis for the project.

This report is intended to cover all the relevant work known to the authors and available in referenced form at the end of 1998. It is not intended to substitute for the original published material, but to contribute to a critical assessment of the state of the art in the field. Relevant but unpublished information was also provided directly by the funding organisations and by other research organisations. In those cases, information is quoted as personal communication, with the agreement of the source. In addition, this report does not provide an ultimate series of recommendations on how to carry out porewater extractions in argillaceous media, but stresses the confidence limits that may be attributed to the results obtained through the different techniques and shows the areas where further investigation is needed.

PART I

INTRODUCTION TO THE CLAY/WATER SYSTEM

1. CLAY MINERALS

It is not the purpose of this study to provide a detailed description of the mineralogy and structure of the clay minerals. The interested reader will find comprehensive treatises in the mineralogy literature (Bayley, 1988; Brindley and Brown, 1980; Brown *et al.*, 1978; Dixon and Weed, 1977; Nemecz, 1981; Velde, 1992). Besides, a detailed summary of the relevant issues may be found in the first report of that series (Horseman *et al.*, 1996). However, a short summary of the main chemical and structural characteristics is necessary for the comprehension of the mechanisms of water and solute interactions with solid phases. Those mechanisms will in fact play a prevailing role during the liquid-extraction procedure.

1.1 Definition and structure

The word "clay" has been generally used for indicating fine-grained sediments (particle size less than 2 μm) with particular attributes of plasticity (Guggenheim and Martin, 1995). That granulometric definition of clays does not take into account the mineralogy of the particles. It appears from X-ray diffraction patterns that clays are mainly constituted, among other minerals such as carbonates, silicates and oxi-hydroxides, by a specific group of aluminosilicates called "clay minerals".

Clay minerals are aluminosilicates consisting of alternating tetrahedral (T) and octahedral (O) sheet structures.

Sheets co-ordinating four oxygen atoms are formed by silicon in a tetrahedral arrangement. The basal oxygen atoms are shared between adjacent tetrahedra as shown in Figure 1. The basic unit formula is therefore $Si_2O_5^{2-}$. Rings of tetrahedra linked together form an hexagonal pattern, named "siloxane cavity", whose form and electric charge is very important in determining the clay properties, as shown later in this report.

Similarly, octahedral sheets are formed by a cation co-ordinating six negative ions (oxygen or OH) and sharing some of those with adjacent octahedra (Figure 1). Two basic types of octahedral sheets may be distinguished: if the central cation is trivalent (*e.g.,* Al^{3+} or Fe^{3+}), only two-thirds of the octahedral sites are occupied, originating the so-called dioctahedral sheet (basic formula $Al_2(OH)_4^{2+}$). If the cation is bivalent (*e.g.,* Mg^{2+}), all the octahedral sites may be occupied, giving rise to the trioctahedral structure ($Mg_3(OH)_4^{2+}$).

Clay minerals are formed by more or less organised alternations of those basic sheets, bound together by electrostatic forces. The origin of the electric charge on the clay surface is investigated in a

Figure 1. Clay structures: tetrahedral (SiO_4) and octahedral (*e.g.*, $Al(OH)_6$) sheets (after Sposito, 1984)

TETRAHEDRAL SHEET

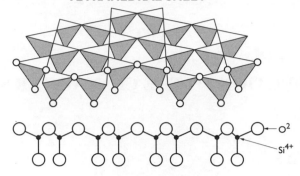

DIOCTAHEDRAL SHEET

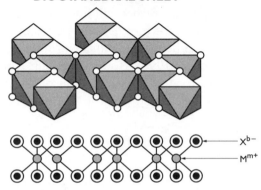

number of studies on colloid chemistry and properties. Besides, cation substitutions in the basic units of the sheets also give rise to an electrical unbalance. For example, silicon in tetrahedra may be substituted by aluminium, leaving an unbalanced negative charge. Any unbalance will be compensated by attracting a positively charged layer or ion close to the surface. The most important clay mineral groups are shown in Figure 2 and are here briefly described.

The kaolinite group minerals display a simple alternation of tetrahedral and octahedral sheets (TO structure or 1:1-layer silicates). In that group, cation substitutions in the basic sheet are not very common, and kaolinites show an almost stoichiometric formula of:

$$Al_2(OH)_4^{2+} + Si_2O_5^{2-} = Al_2Si_2O_5(OH)_4$$

If magnesium is present instead of aluminium, a trioctahedral series is obtained, namely the serpentine subgroup, whose formula is $Mg_3Si_2O_5(OH)_4$. Both those subgroups show an interlamellar spacing (*i.e.,* the distance between two TO groups), of approximately 0.7 nm.

A second group of clay minerals, both di- and trioctahedral, consists of alternating TOT layers (or 2:1 structure), where a substitution of aluminium in the tetrahedral layer is possible. That leaves a negative unbalanced charge that is compensated by the adsorption, between the lamellae, of cations such as K^+. The basic formula of those minerals, named illites and micas, is therefore:

$$K_{(1-X)}Al_2 \{Al_{(1-X)}Si_{(3+X)}\} O_{10}(OH)_2$$

Figure 2. **Clay structures: TO, TOT and TOT O alternations (modified after Millot, 1964)**

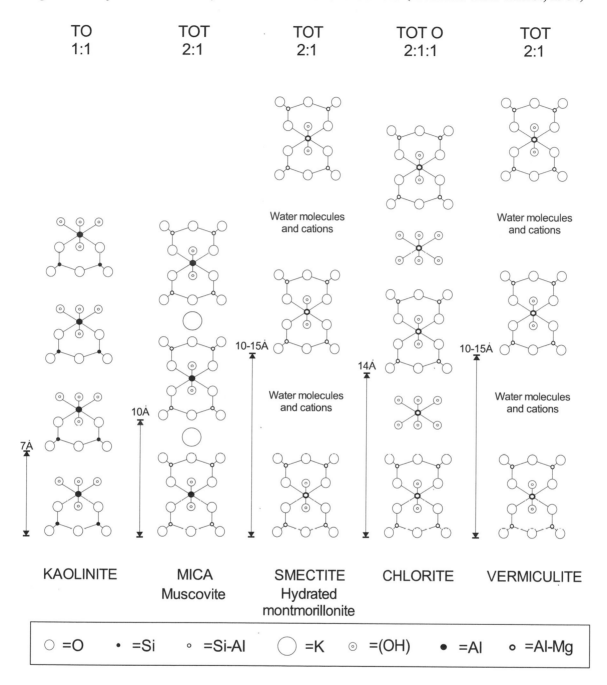

The interlamellar distance is approximately 1 nm.

The smectite and vermiculite groups are very similar, but usually adsorbing different more or less hydrated cations. The interlamellar spacing is variable, depending on the adsorbed species. That swelling property according to the degree of hydration is very important, as we will see later. Smectites and vermiculites are basically distinguished by their permanent structural charge, the latter having a greater one.

Finally, the chlorite group of clay minerals displays a TOT O TOT structure (also called "2:1:1 structure"). The excess of negative charges is balanced by a positive charge on the interlayer hydroxide sheet. That group is highly variable in chemical composition, depending on the possible cationic substitutions in the octahedral and tetrahedral sheet. The structure thickness is approximately 1.4 nm and displays no swelling.

Other minerals such as talc, pyrophyllite, sepiolite and palygorskite belong to the clay-mineral group, but will not be described in detail because of their limited occurrence in the clay environments of interest. The complete clay classification is summarised in Table 1.

Those end-members are not readily found in nature as pure and well-crystallised structures. Clays known as "mixed-layer type" are common; they alternate either regularly or irregularly different sequences of structures. Local changes of structures, tetrahedral and octahedral-cation substitutions and interlayer adsorption are extremely common. Nevertheless, those physical mixtures may be identified by using a number of techniques, including elemental analysis, X-ray diffraction, Infra-Red Spectroscopy (IRS), thermogravimetric analysis, and electronic microscopy. Some authors (Aja *et al.*, 1991; Garrels, 1984; May *et al.*, 1986) treat them as separate, well-defined phases, for which thermodynamic constants are hard to determine, while others prefer to treat them as solid solutions of the described end-members (Fritz, 1975).

1.2   Chemical properties

Most of the properties displayed by clay minerals are due to their small grain size and their sheet-like shape. Both factors cause a very high surface area relative to the mass of material (Van Olphen, 1963).

All clays attract water to their surface (adsorption) and some of them include it in their structure (absorption). A possible classification of clays is based on the way they absorb water (see also Table 1). Smectites are swelling clays, as they increase volume when incorporating water molecules. Other clays, like sepiolite and palygorskite, whose habit is more needle-like, have important sorption capacities, but not the property of swelling. A third group of clays, including kaolinite and chlorite, has neither of those properties.

A related important feature concerns the way clays develop electric charge on the surface. Two main mechanisms are responsible: isomorphic substitution among ions of different charge within the layers and deprotonation of silanol and aluminol groups.

According to the surface charge, clays may be classified in:

– Neutral-lattice structures, both 1:1 and 2:1 where linked tetrahedra and octahedra have a net charge of 0. The substitutions within the sheets are cancelled electrostatically, and layers are bound in the crystal by van-der-Waals-type forces. Talc, pyrophyllite, kaolinite, serpentine and chlorite have that type of structure.

– High-charge structures (0.9-1.0) are observed for micas. The charge unbalance is due to ionic substitutions in the tetrahedral and octahedral sheet, and is balanced by a strong adsorption of cations (mainly K^+) to the surface.

– Low-charge structures (0.2-0.9), where the imbalance is compensated by weakly held ions in the interlayer position that may be readily exchanged in aqueous solution

Table 1. **Clay structure parameters (after Decarreau, 1990 and Velde, 1992)**

Group	Structure	Octahedral layer	Permanent structural charge	Swelling/ Sorbing[3]	Interlayer spacing	Examples
Kaolinite	1:1	Dioctahedral	~ 0	None	7	Kaolinite, dickite, nacrite, halloysite
Serpentine	1:1	Trioctahedral	~ 0	None	7	Chrysotile, antigorite, lizardite, amesite
Pyrophyllite	2:1	Dioctahedral	~ 0	None		Pyrophyllite
Talc	2:1	Trioctahedral	~ 0	None		Talc, willemseite
Mica	2:1	Dioctahedral	~ 1	None	10	Muscovite, paragonite
	2:1	Trioctahedral	~ 1	None	10	Phlogopite, biotite, lepidolitc
Hard mica	2:1	Dioctahedral	~ 2	None	10	Margarite
	2:1	Trioctahedral	~ 2	None	10	Clintonite, anandite
Chlorite	2:1:1	Dioctahedral	Variable	None	14	Donbassite
	2:1:1	Di-trioctahedral	Variable	None	14	Cookeite, sudoite
	2:1:1	Trioctahedral	Variable	None	14	Clinochlore, chamosite, nimite
Smectite	2:1	Dioctahedral	~ 0.2 - 0.6	SW	Variable	Montmorillonite, beidellite, nontronitc
	2:1	Trioctahedral	~ 0.2 - 0.6	SW	Variable	Saponite, hectorite, sauconite
Vermiculite	2:1	Dioctahedral	~ 0.6 - 0.9	SW	Variable	Dioctahedral vermiculite
	2:1	Trioctahedral	~ 0.6 - 0.9	SW	Variable	Trioctahedral vermiculite
Palygorskite	2:1			Variable	SO	Palygorskite
Sepiolite	2:1			Variable	SO	Sepiolite

(*e.g.,* smectites and vermiculites). That type of clays swells, incorporating ions, complexes and molecules between the layers. The interlayer spacing will be determined by the hydration state and the type of ion adsorbed between layers.

3.　　　　SW: swelling; SO: sorbing.

31

The amount of charge per kilogram ($mol_c \cdot kg^{-1}$) created by isomorphic substitutions is called permanent structural charge (σ_S). It may be calculated from the layer charge (x) and the relative molecular mass (M_r) of the mineral:

$$\sigma_S = -(x/M_r) \cdot 10^3$$

σ_S ranges between -0.7 and -1.7 $mol_c \cdot kg^{-1}$ for smectites, -1.9 and -2.8 $mol_c \cdot kg^{-1}$ for illites and -1.6 and -2.5 $mol_c \cdot kg^{-1}$ for vermiculites.

Another property is the cation selectivity (Thomas, 1977). When a solution and clay come into contact and are allowed to establish an equilibrium, some ions will be sorbed preferably to others. In general, divalent ions are preferred in swelling-clay interlayers over monovalent ions, while the opposite may be true for some high charge clays. Cation partitioning depends on many parameters, such as the charge unbalance with respect to the ion charge, the interlayer spacing or the dimension of the adsorbing site (surface functional group) with respect to the ionic radius, the composition of the solution and temperature. As clay particles normally develop a negative charge, anions will tend to be repulsed. That phenomenon is known as "anion exclusion". Anion retention, especially for phosphate, fluorine and sulphate ions, may be observed, and is believed to occur on particular adsorption sites. In addition, natural organic matter shows a strong affinity for clay minerals (Theng, 1974; Maurice *et al.*, 1998).

2. INTERACTIONS BETWEEN WATER, SOLUTES AND CLAY

2.1 Water-molecule structure and cation hydration

The water molecule owes most of its solvent efficiency to its polarity. In fact, the arrangement of the electronic orbitals and the bonds with hydrogen is almost tetrahedral. The angle between the two hydrogen atoms is slightly greater than a pure tetrahedral angle, due to the strong repulsion of the hydrogen atoms. On the opposite side, the two lone pair of electrons form a negative end of the molecule (Figure 3a). In order to minimise the energy of the dipole, the water molecule tends to rotate and create bonds with other ions and molecules. That behaviour is responsible for the excellent solvent capacity of water.

The "structure" of liquid water is not known in detail, but it seems that it is mainly made of small clusters of hydrogen-bonded molecules whose lifetime is approximately 10^{-11} s. At room temperature, an average cluster contains about 40 molecules. With decreasing temperatures, molecules tend to arrange in a structure resembling that of tridymite, a silica polymorph, known as the "ice structure" (an hexagonal network structure, Figure 3b)).

Cation hydration has been deeply investigated (see for example Taube, 1954; Conway, 1981; Marcus, 1985; Burgess, 1988; Neilson and Enderby, 1989; Friedman 1985; Franks, 1985, Güven, 1992 and the review by Ohtaki and Radnai, 1993), and only the most relevant facts to our study will be reported here.

The hydration complexes formed by ions in solution are defined by:

- the distance between the ion and the water molecule;

- the distance between the ion and the protons of the water molecule;

- the time-averaged number of water molecules around the ion in the inner hydration shell (also known as "mean co-ordination number");

–   the mean tilt angles formed by the water molecule to minimise energy.

Neutron diffraction and X-ray powder diffraction may be used to determine those parameters for salt solid samples. The co-ordination number (CN) only gives an indication of the number of water molecules in contact with the cation, regardless of their true interaction with it. The hydration number (HN) is the number of water molecules electrically influenced by the presence of the ion. Those two numbers may differ significantly according to the ratio of the cation radius and its charge. For monovalent ions, CN is greater than HN. For trivalent ions, HN may be greater than CN, suggesting that those ions strongly interact with water molecules beyond their first hydration shell. CN for some ions may vary inversely with the ionic strength of the solution.

Figure 3. **(a) Structure of the water molecule (after Hochella and White, 1990)**
(b) Molecular structure of adsorbed water in the interlayers of halloysite
(kaolinite group)(after Sposito, 1984)

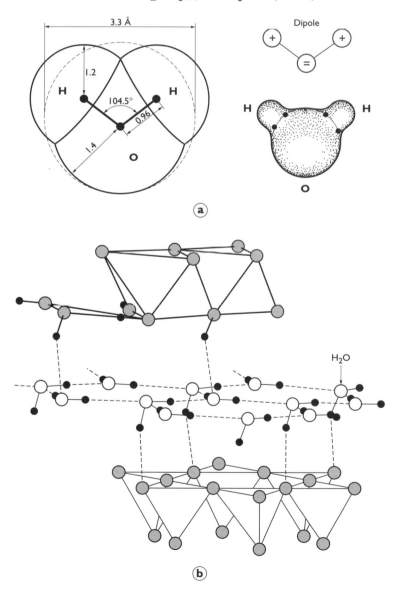

Studies on the mobility of water molecules in the hydration complex have been conducted by spectroscopic methods having the same timescales as the lifetime of those hydration complexes. The stability of the hydration complex may be measured by the exchange rate and/or the mobility of the water molecule in the hydration shell of the ion. The reciprocal value of the exchange rate gives the mean residence time (τ) of a water molecule in the first co-ordination shell of the ion. Those may vary from picoseconds to days. Güven (1992) compiled the mean lifetimes of different hydration complexes and compared them with the lifetime of the water molecule in the liquid and solid phases. He pointed out that hydration complexes with mean residence time close to that of ice are to be considered as rather stable (Figure 4).

Figure 4. **Mean lifetimes of different hydration complexes (after Güven, 1992)**

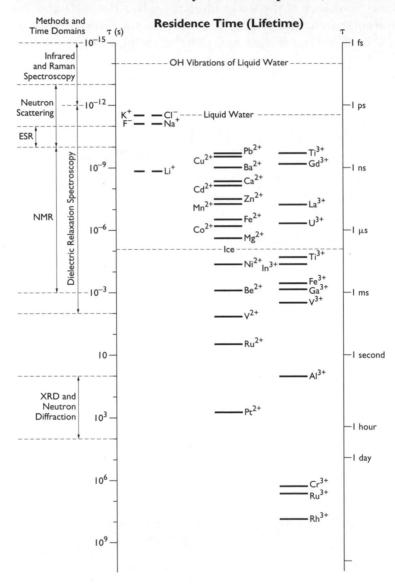

2.2   Clay/water interaction

Some solid particle surfaces tend to develop electric charges when in contact with a liquid phase. As a consequence, the structure of the water molecules in the neighbouring region may be disturbed and differ from that of bulk water (Anderson and Low, 1958; Eger et al., 1979; Cases and François, 1982; Fripiat et al., 1984). The region in which water molecule arrangements differ from that of bulk water is defined as the region of adsorbed water.

2.2.1 Properties of adsorbed water

Low (1982) reviewed several thermodynamic properties of clay/water systems and qualitatively derives that the hydrogen bonds established between the solid and water are more extensible and compressible, but also less deformable than the hydrogen bonds in the liquid structure. Thus, the thermodynamic, hydrodynamic and spectroscopic properties of water in clays differ from those of pure water (Sun et al., 1986). Low (1982) showed, for any thermodynamic property J_i near the clay, that:

$$J_i = J_i^0 \exp \beta \frac{m_m}{m_w}$$

where J_i^0 is the same property of pure water, β is a constant and m_w and m_m are the mass of water and the mass of clay respectively. That relationship seems to hold for high water content media, but cannot be extrapolated to very low water content systems (Sposito and Prost, 1982).

2.2.2 Potential energy of soil and clay water

Water in contact with a solid phase, if in non-equilibrium conditions, will tend to move from a given point in a direction resulting from the combined effects of gravity, hydrostatic pressure and other possible forces, towards a position of lower energy. A number of potentials (pressure, matrix, osmotic, gravitational) of soil water are defined, according to all the possible forces acting on the water (Marshall and Holmes, 1979).

The combination of matrix and osmotic potential is called the "water potential". The matrix potential arises from the interaction of water with the solid particles in which it is embedded (capillary and surface adsorption). The osmotic potential is due to the presence of solutes in porewater. The water potential is measured from the vapour pressure of porewater (e); since it is adsorbed by the matrix and contains solutes, it is lower than that of pure, free water at the same temperature (e_0). The quantity of water on which potentials are based may be a mass, a volume or a weight. The water potential expressed per unit mass (in joules per kilogram, $J \cdot kg^{-1}$) is equal to:

$$\text{water potential} = RTM^{-1}\ln (e/e_0)$$

where R is the gas constant (8.3143 $J \cdot K^{-1} \cdot mol^{-1}$), T is the temperature in kelvins, M is the mass in kilograms of 1 mole of water (0.018015).

Potentials may also be expressed per unit volume in newtons per square metres ($N \cdot m^2$) or pascals (Pa), or per unit weight as lengths (head) in metres. The bar (10^2 kPa), whose magnitude is about 1 atmosphere, and the unit pF, equivalent to the decimal logarithm of the suction expressed in centimetres, have been used in the past, and will be sometimes used in this study, although they are not part of the International System. The values of the matrix potential in different situations, expressed in those units, are indicated in Table 2.

Table 2. **Matrix potential in different units for soil water
(modified after Marshall and Holmes, 1979)**

Conditions at the quoted potential	Matrix potential				
	Per unit volume		*Per unit mass*	*Per unit weight*	
	kPa	*bar*	*J·kg⁻¹*	*m*	*pF*
Saturated or nearly saturated	10^{-1}	10^{-3}	10^{-1}	10^{-2}	0.0
Near field capacity	10	10^{-1}	10	1	2.0
Near permanent wilting point	$1.5 \cdot 10^3$	15	$1.5 \cdot 10^3$	$1.5 \cdot 10^2$	4.2
Air dry at relative vapour pressure of 0.48	10^5	10^3	10^5	10^4	6.0
Conversion from a matrix potential of h metres	$9.8\,h$	$0.098\,h$	$9.8\,h$	h	$\log 100\,h$

When the influence of soluble salts may be neglected, the difference in water vapour pressure e/e_0 may be used to calculate the matrix potential through the relationship:

$$\psi = RTM^{-1}\ln (e/e_0)$$

Figure 5 shows the curves displayed by various clay minerals when reporting the water content *versus* the relative vapour pressure (or the matrix potential). As a reference, for values of e/e_0 higher than 0.989 corresponding to a matrix potential of -1.5 MPa, soil is moist enough to support plant life, and water shows the same properties as bulk water.

Figure 5. **Adsorption of water vapour by different clays (after Marshall and Holmes, 1979).
Considering that the density of water is close to 1,000 kg·m⁻³, values expressed
per unit mass (kJ·kg⁻¹) or per unit volume (kPa) are approximately equal**

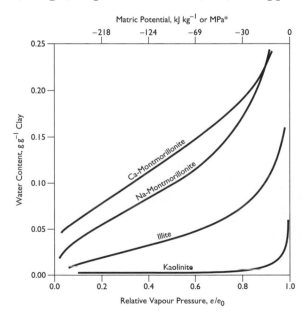

The notion of water potential, and its measurement, has been developed for systems not saturated with water. In our case studies, *a priori*, the deep clays are fully saturated, but the process of water extraction involves in most cases the desaturation of the sample. That concept has been introduced, to show, as we will see later, that the total amount of water is not the prevailing parameter, but rather the strength with which the water is attracted to the solid surface. It is consequently assumed that the same type and entity of forces between clay and water are acting in both the unsaturated and in the saturated medium. A detailed discussion on the validity of that assumption may be found in Horseman *et al.* (1996).

2.2.3   *Types of water in the clay/water system*

Clay hydration involves adsorption of a number of water molecules on the exposed surfaces of clay particles. As a consequence, the specific surface area will influence the amount of adsorbed water. Smectite, with its expanding crystal lattice, adsorbs much more water at a given value e/e_0 than kaolinite, displaying larger crystals and a smaller specific surface area available for adsorption (Figure 5). Illite has an intermediate position.

Three adsorption processes seem to take place with the increase in water content (Figure 6):

a) Adsorption of water in the interlamellar space, inside clay particles (Type I), named "interlamellar water" (see also Figure 2);

b) Continuous hydration related to unlimited adsorption of water around primary clay particles (Type II), named "intraparticle water";

c) Free-water condensation in micropores (Type III), named "interparticle water".

Figure 6. **Different types of water (modified after Allen *et al.*, 1988)**

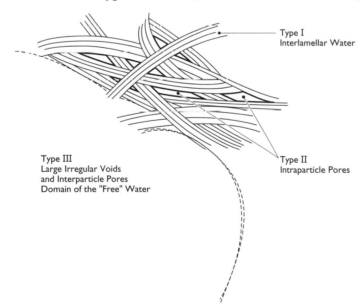

In a water-content *versus* pF plot (Figure 7), the three above-mentioned types may be distinguished. Each of the adsorption stages involves different types of forces and water movement mechanisms (Güven, 1992). The first two fields strongly relate to the clay structure itself, and we will examine them in more detail.

Figure 7. **Water adsorption isotherms (after Decarreau, 1990)**

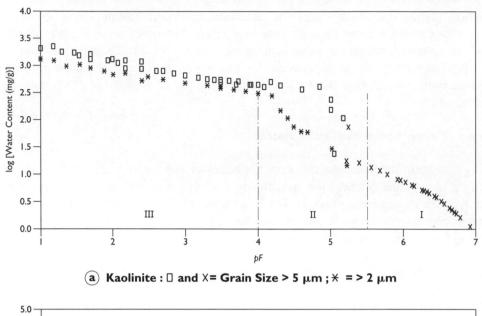

ⓐ Kaolinite : □ and X= Grain Size > 5 μm ; ✳ = > 2 μm

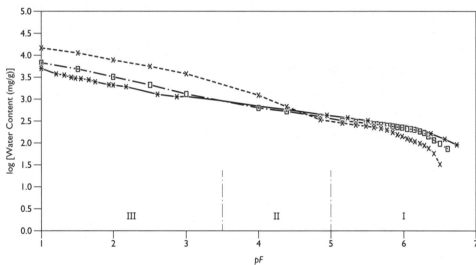

ⓑ X = Hectorite ; □ = Ca-Montmorillonite ; ✳ = Na-Montmorillonite

In the first region, corresponding to very low water contents, adsorption occurs on the hydrophilic sites of the clay surface. Kaolinite-group minerals develop low surface charges, and therefore only adsorb up to a monolayer of water molecules in the interlayers. A hydrogen bond is first established between the water molecules and the basal oxygen atoms of the tetrahedral layer. At higher water contents, the monolayer of water molecules is formed and bonds may be established between the water molecules (Figure 3b). Given the constraints of the silicate structure, bonds have to adjust by stretching and rotating in order to minimise their energy. That results in a configuration that resembles that of bulk liquid water, as a compromise between the disorder of the liquid phase and the rigid arrangement imposed by the solid structure. Motion of the water molecules is consequently slowed down because of the enhancement of the hydrogen bonding.

In 2:1-type clays, the presence of a cation adsorbed on the surface to balance electric charge deficiencies is extremely important (Anderson and Low, 1958). In fact, it appears that the first stage of hydration proceeds through cation hydration. Water molecules are co-ordinated in a single solvation shell for monovalent cations and in a double shell for divalent cations. Hydrated vermiculites contain interlayer ionic solutions (De La Calle *et al.*, 1985). Smectites hydrate in a very similar way: water molecules tend to arrange around cations adsorbed in the siloxane cavities, and subsequently form a monolayer with a strained, ice-like structure.

Prost (1975) conducted an extensive study on the smectite hydration using IRS on progressively deuteriated samples. He examined both the arrangement of the water molecules around the clay particle and the disturbance induced on the clay structure by the presence of a water molecule. He distinguished three types of water molecules with different symmetries and links: those solvating cations, those present around the clay particles and those not in contact with the solid. At very low water content (less than 10% weight, approximately equal to a water monolayer between clay layers) smectites only display the first two types of water molecules. Adsorbed water molecules show a different arrangement if the charge deficiency originated in the octahedral or tetrahedral layer. In the first case, as the permanent structural charge is lower and diffuse, one of the hydrogen atoms of the water molecule is oriented towards the centre of the siloxane cavity, thus disturbing their OH functional groups (see later discussion). In the second case, hydrogen bonds are established between the water molecule and the oxygen atom on the surface of the layer.

Another mechanism that seems to occur in the first hydration stage is capillary condensation at the contact points between particles and/or grains. Capillary condensation will continue and constitutes the main adsorption mechanism in the previously defined second domain. That gives rise to Type-II water in the interparticle pores. That water is still bound to the clay particles and forms monolayers of water molecules, whose thickness is proportional to the water activity.

For values of e/e_0 corresponding to pF < 3.5 for montmorillonites and 4 for kaolinites, we are in the third domain, corresponding approximately to the field of free water (Type III).

It is worth noticing that adsorption and desorption of water onto clays are often affected by hysteresis phenomena (Sposito and Prost, 1982). It is impossible to extrapolate information derived from an adsorption isotherm to water desorption. That observation is normally justified in soils by the so-called "ink-bottle" effect, (Marshall and Holmes, 1979) a capillary effect arising from the difference in diameter of the pores and their openings. In clays, hysteresis is observed even if the clay remains saturated during the cycle, and a capillary effect cannot be held responsible for that observation. A possible explanation relies on particle rearrangement and pore-distribution changes during swelling and shrinking. That phenomenon has been extensively studied (Boek *et al.*, 1995) and must be remembered because of its possible consequences on the isotopic composition of the interstitial solutions.

2.3   Clay/solute interaction

Reactions between the solid and the solution involving mass transfer from the latter to the former may be of three different types: adsorption, absorption and precipitation. Adsorption is defined as the process through which a net accumulation of a substance occurs at the common boundary of two phases (Sposito, 1984). It is essentially a two-dimensional process. Absorption corresponds to the diffusion of the adsorbed species into the solid-phase crystal lattice. Precipitation leads to the formation of a new bulk solid phase and is a three-dimensional process. All processes imply the loss of material from an aqueous solution phase, and often may not be very clearly distinguished, especially in natural systems. In fact, the chemical bonds involved are sometimes similar and mixed precipitates may display a component restricted to the surface because of poor diffusion. When no specific data are available, the generic term "sorption" should be used. Laboratory studies often concentrate on a single type of clay and the reactions may be more readily identified. Cations adsorbed as inner-sphere complexes are commonly designated as "specifically sorbed" or "chemi-sorbed". Monovalent cations usually display differences in the selectivity, mainly reflecting the formation of inner-sphere complexes. Each monovalent cation is settled into a siloxane ditrigonal cavity. Experimental differences may be related to the facility with which a monovalent ion may loose its hydration shell to fit into the siloxane cavity. Ions found in outer-sphere complexes or in the diffused ion swarm are non-specifically sorbed. That is the case of most divalent cations, although it is proven that divalent cations may also form inner-sphere complexes. The distinction also relates to the strength of the bonds linking the ions to the clay surfaces: outer-sphere ions exchange more readily than inner-sphere ions.

2.3.1 *Sorption mechanisms*

The effect of cation hydration is very important for the structure and properties of hydrated clays, especially regarding sorption by compensation of structural charge unbalance: cations with a low hydration energy, such as K^+, NH_4^+, Rb^+ and Cs^+ produce interlayer dehydration and layer collapse, and are therefore fixed in the interlayer positions. Conversely, cations with high hydration energy such as Ca^{2+}, Mg^{2+} and Sr^{2+} produce expanded layers and are more readily exchanged (Sawhney, 1972). That difference may be also regarded as the tendency of a given cation to form "inner sphere" or "outer sphere" complexes (see next paragraph).

A second sorption mechanism on clays may be described in terms of reactions between dissolved solutes and surface functional groups. The latter are reactive molecular units that protrude from the solid adsorbent into the liquid (Sposito, 1984). Analytical methods for their investigation are described in Davis and Hayes (1986).

Clays display proton-bearing surface functional groups (Brönsted acids like OH). Hence, adsorption on those sites is pH-dependent. Different types of surface hydroxyl groups may be distinguished, with different reactivities, depending on the co-ordination environment of the oxygen (Figure 8). Those are designated A, B, or C depending on whether the oxygen is co-ordinated with 1, 3 or 2 cations, respectively. A sites are amphoteric (*i.e.,* they may act as proton donors or else as basic sites forming a complex with H^+). B and C type hydroxyls are considered as non-reactive, but may be turned into A type, depending on the pH of the solution.

A third type of sorption site results from the Lewis acidity of the metallic cation, enabling it to exchange an OH^- ligand with a stronger Lewis base, such as F^-, for example (Mortland and Raman, 1968).

Densities of surface hydroxyl groups may be measured by IRS, isotopic exchange, thermogravimetry and reaction with OH labile compounds (James and Parks, 1982), while densities of proton donors and acceptors are calculated from crystallographic considerations (Sposito, 1984).

The plane of oxygen atoms bounding a tetrahedral silica sheet in a silicate layer is called a "siloxane surface". The functional group associated with the siloxane surface is called the "siloxane cavity", formed by six corner-sharing silica tetrahedra. Thus, phyllosilicate minerals display, in addition to the previously described surface functional groups, rings of siloxane group located on the tetrahedral basal plane. Linkage of the tetrahedral to the octahedral plane causes the distortion of the cavity from hexagonal to ditrigonal. The reactivity of the siloxane cavity depends on the nature of the electronic charge distribution in the phyllosilicate structure (Sposito, 1984). Cation substitutions within the layers create permanent structural charges compensated by the complexation of mono or divalent cations into those cavities.

Figure 8. Surface functional groups: a) Surface hydroxyls and Lewis acid sites on goethite; b) Aluminols and silanols; c) The siloxane cavity (after Sposito, 1984; 1989)

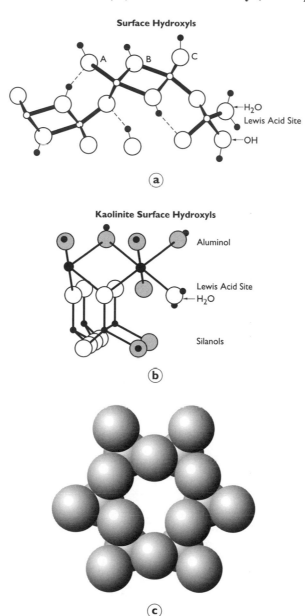

In summary:

– Kaolinite has five types of surface functional groups: (1) the siloxane cavities on the faces of the tetrahedral sheets, (2) the aluminols ($\equiv AlOH^-$) on the faces of the octahedral sheets, and (3) silanols ($\equiv SiOH^-$), (4) aluminols and (5) Lewis acid sites on the edges of the clay crystal. In kaolinite, the degree of ionic substitution in the lattice is very low (< 0.01 ion per unit cell). As a consequence, the siloxane cavities are almost non-reactive. They will

41

therefore act as mild electron donors, binding only neutral and dipole molecules like water. The most important surface complexation sites will be silanols, aluminols and Lewis acid sites on the edges of the crystal. All of them are proton donors and may form complexes with metal ions. Aluminols are also proton acceptors and will be able to complex anions.

– Smectites, vermiculites and illitic micas have a significant permanent charge due to ionic substitutions. If those occur in the octahedral layer, the charge deficiency is distributed on ten surface oxygen atoms of the tetrahedral sheet, belonging to four tetrahedra linked to the site. The charge is diffuse and leads to the formation of outer-sphere complexes (Figure 9). If the substitution is directly in the tetrahedral sheet, the siloxane cavity bears a localised charge deficiency. That enhances the formation of inner-sphere complexes. Since the K^+ radius is very close to the diameter of the siloxane cavity (0.26 nm), it may form a very stable complex.

Figure 9. **Inner sphere and outer sphere complexes (after Sposito, 1984; 1989)**

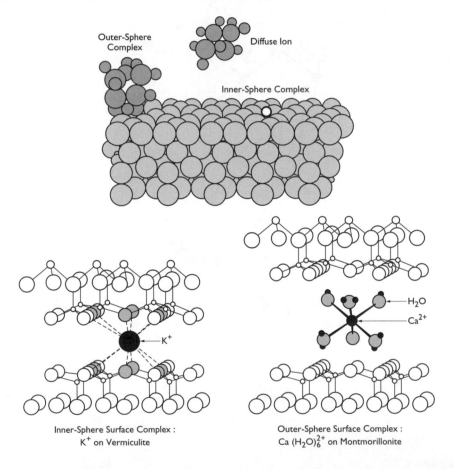

Clay minerals forming outer-sphere complexes have readily accessible siloxane cavities: as a consequence, those cations are readily exchangeable. The clay in contact with the solution will tend to react, thus populating those sites with the dominant cation of the solution (generally Ca^{2+}, regarding abundance and affinity). Anion binding on clays occurs predominantly along the edges of the crystals.

2.3.2   *Electric double layer*

As we have already seen, clays develop an electric charge when in contact with a solution. The net total surface particle charge σ_P [C·m^{-2}] is defined as:

$$\sigma_P = \sigma_S + \sigma_0$$

where σ_S is the permanent structural charge, arising from cation substitutions in the clay lattice, and σ_0 is the co-ordinative surface charge, associated with the reaction of potential-determining ions with the surface functional groups. The co-ordinative surface charge σ_0 may be distinguished according to different contributions:

$$\sigma_P = \sigma_S + \sigma_H + \sigma_{is} + \sigma_{os}$$

σ_H is called the net proton charge: it is the difference between the specific charges of protons and hydroxide ions complexed by the surface functional groups:

$$\sigma_H = q_H - q_{OH}$$

Values for σ_H are measured by titration as a function of the pH of the solution. σ_S and σ_H form the intrinsic surface charges of clays, as they arise mainly from structural components.

The two other terms, σ_{is} and σ_{os} come from the inner and outer sphere complexes respectively. Their sum arise solely from the solution ions adsorbed on the surface (excluding H$^+$ and OH$^-$). They may be measured with specific ion-selective electrodes or by ion displacement methods. They are distinguished based on the established links with the clay mineral: covalent bonding retains inner-sphere complexes, while outer-sphere complexes show only weak electrostatic bonding. For example, K$^+$ in the inner sphere will contribute +1 mol$_c$, and Ca^{2+} in the outer sphere will contribute +2 mol$_c$ to those terms.

While σ_S is almost always negative, σ_H, σ_{is}, σ_{os} and the net particle charge σ_P may be either positive, zero or negative, depending on the surrounding solution composition. If σ_P is not equal to zero, another balancing charge must be accounted for to preserve electroneutrality in the clay/water system. That arises from the ions that are not bound to the surface, but yet adsorbed in the diffusion ion swarm (σ_D). In that case, the sum of σ_P and σ_D is equal to zero.

Gouy (1910) and Chapman (1913) derived equations for describing the distribution of the counterions in the diffuse ion swarm formed by a charged planar surface. The detailed derivation and discussion of the equation is given in Bolt (1982) and Sposito (1984). In the model, all the counterion charge σ_D is present as dissociated charge, and electroneutrality is given by:

$$\sigma_P + \sigma_D = \sigma_0 + \sigma_S + \sigma_D = 0$$

while σ_D for a symmetrical electrolyte with ions of charge z at 25°C is derived from the Poisson-Boltzmann equation:

$$\sigma_D = -0.1174\sqrt{I} \cdot \sinh \frac{ze\psi_0}{2kT}$$

ψ_0 being the electric potential at the surface. Thus, the calculated potential decays exponentially with the distance from the clay particle surface (Figure 10). That relation is valid for symmetrical electrolytes, while for asymmetrical ones, a different charge/potential relationship is involved (Hunter, 1989).

Figure 10. **Schematic drawings of the electric double layer according to the Gouy-Chapman and Stern-Grahame models (after Hochella and White, 1990). The relationships shown assume $\sigma_S = 0$**

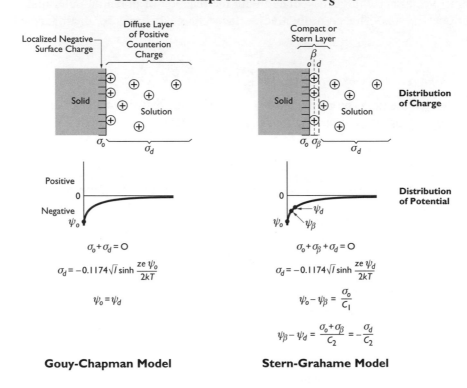

The Gouy-Chapman theory was found to be in poor agreement with experimental data. Stern (1924) and Grahame (1947) introduced a few modifications taking ion size into account. In their model, some ions may be bound to the surface by specific adsorption on a plane very close to the charged surface. As a consequence, the particle surface charge is balanced by the charge in the "Stern layer" (β plane) and the dissociated charge:

$$\sigma_P + \sigma_\beta + \sigma_D = 0$$

The potential decays linearly between the planes of the Stern layer (also known as "inner Helmholtz plane" and "outer Helmholtz plane") and then exponentially, according to the Gouy-Chapman equation (Figure 10).

Points of zero charge are pH values at which some of those contributions to the particle charge turn to zero (Table 3). In particular, at the point of zero charge when σ_P is equal to zero, clay particles do not move in an applied electric field. That may also cause settling and flocculation, enhancing the coagulation effects.

Table 3. **Points of zero charge (modified from Sposito, 1989)**

Definition	Charge properties
pH point of zero charge (pH_{PZC})	$\sigma_P = 0$
pH point of zero net proton charge (pH_{PZNPC})	$\sigma_H = 0$
pH point of zero net charge (pH_{PZNC})	$\sigma_{is} + \sigma_{os} + \sigma_D = 0$
pH isoelectric point (pH_{IEP})	$\sigma_D = 0$

2.3.3 *Dissolution and precipitation*

In the presence of an aggressive fluid (*e.g.,* fresh water), clay minerals may dissolve and the amount of resulting species in the water increase. At a given level of concentration, some of them may combine chemically, and neo-formed phases may precipitate. The saturation effect is responsible for the limitation of the concentration of the *geochemically controlled elements* in natural fluids. That phenomenon is very important for modelling the groundwater chemistry (see Chapter III).

3. POROSITY, SALINITY AND HYDRATION

3.1 High-water content systems

Porosity has been studied extensively, especially during the clay evolution from sedimentary to low metamorphic through diagenesis (see Horseman *et al.*, 1996 for a good review).

In very dilute solutions, clay particles tend to aggregate, in particular conditions, into stacks of roughly-parallel single layers, called "quasi-crystals" (Quirk and Aylmore, 1971). That particle structure is stabilised by the attractive forces between the basal planes of single-layer platelets mediated by adsorbed cations and water (Sposito, 1984). Quasi-crystals appear to form from any bivalent cation and any smectite (Sposito and Prost, 1982; Sposito, 1992): montmorillonite forms quasi-crystals comprising stacks of four to seven layers. Ca^{2+} ions, solvated by six water molecules (outer-sphere complex), serve as molecular cross-links to help the clay layers bind together through electrostatic forces. Large monovalent (K^+, Cs^+) ion-containing smectites may also form quasi-crystals, but interlamellar hydrates may be limited to one or less layers (Güven, 1988). When sodium and lithium interlayer cations are immersed in an aqueous solution, the clay is often dissociated in individual silicate layers, whose hydration complexes consist of a continuous diffuse double layer. Such a set of smectite layers that are separated by their overlapping double layer is called "tactoid". That type of hydration is not limited as for quasi-crystals; it is named "osmotic swelling" and is found to be inversely proportional to the square root of the salt concentration of the solution (Norrish, 1954). The existence of sodium-saturated smectite quasi-crystals in stacks of two single-layer platelets with three layers of water molecules between them has also been indirectly reported (Sposito, 1989).

At very low densities of the clay/water system, there seems to be evidence of an edge/face mode of flake association, while, as density increases, the face/face mode of association is more likely. Water and solutes are then retained as interlamellar solutions (or internal water) or in the voids between stacks (external solutions) (Pusch and Karnland, 1986; Pusch *et al.* 1990). As permeability decreases (*i.e.,* the large pores are compressed and water expelled), the ratio of external to internal water decreases.

Many studies have been conducted on the chemistry of solutions expelled by compacting clays during diagenesis to explain the high salinities found in those formations (Engelhardt and Gaida, 1963; Chilingarian *et al.*, 1973; Rosenbaum, 1976; Lawrence and Gieskes, 1981). Those studies showed a concentration decrease of the expelled solutions with increasing pressure. Compaction rate and clay texture are influenced by the salinity of the solution: compaction would proceed more rapidly and clay platelets would be less arranged in the presence of an electrolyte. Those phenomena are not observed for kaolinite, indicating that the cation exchange capacity of the clay would be responsible for them. A model based on the Donnan principle (Appelo, 1977) was found to describe adequately the expulsion mechanism. The anion concentration should be lower next to the particle surface, and consequently the liquid immediately surrounding a clay particle should contain fewer electrolytes than the free solution. During compaction, the free solution would be removed and the electrolyte-poor solution of the electric double layer left behind. High salinities in the free porewater would prevent clay platelets to come too close to each other, thus increasing the permeability and the disorder of the structures. Compaction would proceed more rapidly because of the higher permeability maintained (Hardcastle and Mitchell, 1974).

In normally compacted zones, shale porosity and porewater salinity are reciprocal. If no salt were lost by fluid expulsion, a unit volume of shale after compaction would contain more salt than before compaction. It is found (Magara, 1974 for the Gulf Coast) that the salt content is approximately the same before and after compaction. As a consequence, calculated salinity of expelled fluid accounts for about one-third of the original salt content. In undercompacted zones, porewater salinities are lower, but the product salinity-porosity decreases, indicating an additional "refreshening" of the solution. In alternating clay/sand sequences, the distribution of porosities and salinities within the clay layers shows the presence of additional mass transport mechanisms, such as osmotic flow (Hall, 1993). Besides, laboratory experiments showed a semi-permeable behaviour of clay membranes with respect to solutes and isotopes (Coplen and Hanshaw, 1973; Hanshaw and Coplen, 1973; Kharaka and Berry, 1973; Kharaka and Smalley, 1976; Charles *et al.*, 1986; Phillips and Bentley, 1987; Demir, 1988). That mechanism would enhance transport across the clay for divalent cation over monovalent in a selectivity sequence approximately related to their ionic radius. We will not review in more detail the flow and transport mechanisms within and across clay formations. Horseman *et al.* (1996) produced a very detailed and updated monograph on that topic.

3.2   Low water-content systems

At the other end of the hydration spectrum, studies have been conducted on dry clays, especially smectites, incorporating water molecules. The interlamellar hydration of clays by a water-vapour phase has been well documented by Hendricks *et al.* (1940), Mooney *et al.* (1952 a; b), Norrish (1954), van Olphen (1965; 1969), McEwan and Wilson (1980), Suquet *et al.* (1975; 1977), Suquet and Pezerat (1987), Kraehenbuehl *et al.* (1987), Kahr *et al.* (1990) and Yormah and Hayes., 1993. The interlamellar-hydration phase generally begins with the formation of primary hydration shells of the interlayer cations, prior to the monolayer coverage of the clay surface.

Prost (1975) combined his observations obtained with IRS on hectorite with structural data and computed, for low water-content systems (< 10%) the percentages of the clay surfaces covered by water and the percentage of water not in contact with the clay as a function of the clay water content. He found that those percentages were very different for the same clay, depending on the type of saturating cation. According to those observations, the hydration energies of cations play a prevailing role in defining the amount of adsorbed water (Figure 11).

Figure 11. **Evolution of the water content of a smectite as a function of the saturating cation (after Decarreau, 1990)**

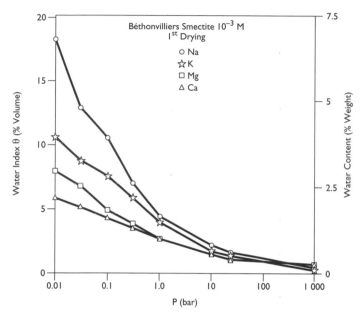

Sposito and Prost (1982) concluded from IRS studies, that solvation of the exchangeable cations either by three (monovalent ions) or more (bivalent ions) water molecules forms the first stage of water adsorption on smectites. As a consequence, interlamellar spacing has to expand in order to accommodate water molecules (Karaborni et al., 1996). Additional hydration results in the formation of solvation complexes or sheaths for the exchangeable cations. Besides, investigations of the dielectric-relaxation properties of montmorillonite saturated with monovalent exchangeable cations indicate that water in the interlamellar space is arranged in a monolayer with an ice-like structure: some of the water molecules are thought to be strongly associated with the oxygen atoms on the silicate surface. Instead, in calcium-saturated montmorillonite, the cation tends to bind the solvation shell strongly, eventually disturbing the water lattice.

Pusch and Karnland (1986) and Pusch et al. (1989, 1990) attempted, based on microstructural analysis, to evaluate the porosity and discriminate between internal (interlamellar) and external water. It is concluded that smectite-rich materials hold an amount of internal water that mostly depend on the bulk density of the clay/water system. Tardy and Touret (1987) found that the water content of smectites increases exponentially as the relative vapour pressure is increased from 0.96 to 1.0. Upon saturation of clay, the microstructure of clays appears to be the main factor in the hydration process. High-resolution transmission electron microscope (TEM) and small-angle neutron-scattering studies by Tessier and Pedro (1987), Ben-Rhaiem et al. (1987) and Touret et al. (1990) also documented the water partition over three kinds of pores (interaggregate, intra-aggregate and interlamellar) (Table 4). They concluded that in saturated clays most of the water seems to occur in the inter and intra-aggregate pores, whose dimensions are determined by the morphologic features of smectite particles, such as lateral extension and flexibility of smectite films. On the other hand, since an increase of the ionic strength of the solution produces coagulation of the particles, for any particular porewater salinity there is a unique relationship between internal and external water (Pusch and Karnland, 1986) (Figure 12).

Figure 12. **Theoretical relationship between dry bulk density and content of "internal" water expressed in percentage of the total pore volume (after Pusch *et al.*, 1990)**

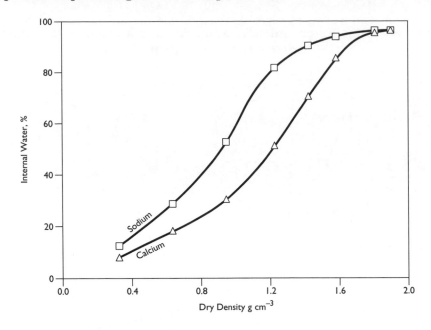

In summary, the position, size and charge of interlamellar exchangeable cations, that in turn depend on the location of the deficit of positive clay-lattice charge, largely determine the spatial arrangement of water molecules. Recent studies on the water mobility in clay-rich systems support the notion that water is more influenced by the saturating cation than by the particular clay (Weiss and Gerasimowicz, 1996).

Table 4. **Water distribution over three kinds of pores for different clay minerals (modified after Touret *et al.*, 1990)**

Clay mineral	Interparticle pores	Intraparticle pores	Interlamellar space	Total amount of water per gram of clay
	% of total amount of water per gram of clay			
Hectorite	25	40	35	1.46
Montmorillonite (Wyoming)	33	42	25	1.79
Montmorillonite (Camp-Bertaux)	58	20	22	2.08
Nontronite	52	24	24	1.72
Vermiculite	69	12	19	1.12

3.3   Chemical porosity

A few interesting studies relate to pore-size distribution and to pore accessibility to solutions. Unless organic hydrophobic matter coats the pore walls, water usually wets their surface. As the hydration boundary may extend as far as 10 nm (Yariv and Cross, 1979), the degree of water "immobilisation" is greater near the pore walls than in the centre of the pore. If pores are less than a few tens of nanometres in diameter, the water and the solutes cannot flow unless a threshold pressure gradient is exceeded. In addition, because of the negative charges developed by the clay mineral surfaces, some regions of the pores may be precluded access to negative ions.

As a consequence, different types of porosity may be defined (Pearson, in press):

- Physical porosity is the ratio of void volume to total volume. For fully saturated clays, total physical porosity (n_t) may be calculated from the dry bulk density ρ_b and the average grain density of the mineral solids ρ_s:

$$n_t = 1 - \frac{\rho_b}{\rho_s}$$

That porosity includes isolated pores and fluid inclusions.

- Water content porosity (n_{wc}), describing the connected, rather than the total, porosity is determined by the difference in weight between the dry sample and the water-saturated sample. It is important to notice that, for mudrocks, the amount of water loss depends on the drying conditions, which will in turn determine the amount of interlamellar water removed from the sample. In general, the total physical porosity is greater or equal to the water-content porosity:

$$n_{wc} \leq n_t$$

- Transport porosities refer to the velocity of a substance in a fluid. As a consequence, they are not only a property of the rock but also of the substance being transported. Those porosities are derived from tracing experiments and may be distinguished, based on the main transport mechanisms, as advection porosity (n_{adv}) and diffusion porosity (n_{diff}). For its definition, advection porosity does not take into account isolated pores and pores opened only at one end. In mudrocks, it is normally found, for the water molecule itself that:

$$n_{adv} \leq n_{diff} \approx n_{wc} \leq n_t$$

while for any other substance:

$$n_{adv} < n_{diff} < n_{wc} \leq n_t$$

- Geochemical porosity (Pearson, 1998; in press) represents the fluid volume in which reactions occur. It is required for geochemical and reactive transport modelling. It is similar to transport porosities, and in clay-rich materials is closer to diffusion porosity.

In coarse-grained rocks, all porosity types are approximately equal because of the lack or the minor influence of attractive and repulsive forces exerted by the solid phase. Pearson (1998) reported estimates of those different porosity types for clay-rich rocks (London clay, clay-rich Canadian tills, Boom clay, Opalinus clay and Palfris marl). According to those calculations, geochemical and diffusion porosities for water molecules are the same and are equal to water-content porosity. Geochemical and diffusion porosities for solutes that do not have access to interlayer or surface sorbed waters are only one-third to one-half of the water content porosity. Those values vary with the salinity of the solution, since that is shown to modify the thickness of the double layer and the pore-size distribution due to osmotic swelling (Karnland, 1997). That observation is of primary importance for the calculation of the porewater composition derived from leaching experiments.

4.   ORGANIC MATTER

4.1   Definition, origins and composition of the organic matter

Organic matter has been the focus of special attention since its importance as sink for trace metals and radionuclides was established (Sposito, 1984; Buffle, 1984; 1988; Yong *et al.*, 1992). The organic-matter content of a soil or sediment originates from the biological activity hosted by the environment. In soils, humic substances are the major organic constituents, arising from the chemical and biological degradation of plant and animal residues (Schnitzer, 1978). Various types of organic substances may be found in sediments, including aliphatic and aromatic hydrocarbons and non-hydrocarbons geopolymers known as bitumen and kerogen (Yariv and Cross, 1979). The organic-matter characterisation of rocks may provide information on the biological input, the palaeodepositional environment, and the degree of maturity and degradation. In our discussion, we will try to define and distinguish within the organic matter, the soluble fraction, that may be found in porewaters, from the solvent extractable fraction (bitumen) and the non-extractable fraction (kerogen). Those three fractions show different properties and are related in different manners to the main topic of this study.

In surface waters and groundwaters, the total organic carbon (TOC) includes the particulate organic carbon or POC (size > 0.45 µm) and the dissolved organic carbon or DOC (size < 0.45 µm), the latter representing in most natural water systems more than 90% of the total (Thurman, 1985a). Plant or animal debris, bacteria and mineral particles, like clays coated with adsorbed organic substances, make the POC.

Most of the efforts in recent years have been put in the isolation, purification and characterisations of the different forms of DOC. Some components, approximately 20% of the DOC, may be identified by using specific techniques such as gas and liquid chromatography coupled with mass spectrometry or spectrophotometry. However, because of the complexity of their structure, the other fractions may only be defined as a function of the technique used to separate them. Leenheer (1981) proposed analytical procedures to fractionate the bulk DOC, according to which DOC is constituted by:

–   Humic substances including fulvic and humic acids;

–   Non-humic substances, including hydrophilic acids, carbohydrates, carboxyl acids, amino acids and hydrocarbons.

Their proportions vary in different environments: in surface and subsurface waters, humic substances represent 40 to 60% of the DOC, while in deep groundwaters hydrophilic acids account for more than 50% of the DOC. Humic substances are divided in fulvic acids that remain in solution over the whole range of pH, and humic acids that precipitate at low pH (Thurman, 1985b).

The chemical structure of humic substances is not completely defined, even if a number of models are proposed in the literature (Schnitzer, 1978; Hayes, 1985). They may be regarded as three-dimensional polymers with a high molecular mass, and a more or less aromatic character. Schematically, they are constituted by an aromatic polycyclic nucleus at which lateral chains of proteins or polypeptides are fixed through amino-acid bindings. There is no clear structural separation between fulvic and humic acids, nor a clear idea of their relationship (*i.e.,* whether fulvic acids are the precursors of humic acids or vice versa) (Andreux and Munier-Lamy, 1994). Hydrosoluble acids may be regarded as low molecular mass compounds of the hydrophilic fraction, in contrast to humic substances showing a higher molecular mass. The difficulties met in their isolation and purification in a sufficient amount has prevented up to now a reliable study of their behaviour with respect to radionuclides. However, a recently developed extraction method (Dierckx *et al.*, 1996; Devol-Brown *et al.*, 1998) offers interesting new perspectives.

Two main origins for the DOC are possible (Thurman, 1985b). A pedogenetic origin leads to the formation of humic substances via the decomposition of plant and animal debris by micro-organisms that transform them into sugars, polyphenols and modified lignin (humification). The remaining part is the lignin itself. Those four types of compounds, in combination with amino compounds, form humic substances through four different reaction pathways (Felbeck, 1971; Schnitzer, 1978; Beaufays *et al.*, 1994), that account for the variety of the final product. Water percolating through the soil is then responsible for the transport of soluble fractions into the aquifer. A sedimentary origin of the DOC is also possible. Recent sediments contain organic products very close to those that may be found in soils, corresponding to the decomposition of plant and animal debris. Organic matter tends to evolve during diagenesis, catagenesis and metagenesis, forming kerogens and oils (Figure 13). The ultimate stage is reached during metagenesis with methane formation (see Yariv and Cross 1979, for a more detailed summary). If, later, the sediment becomes an aquifer, the water will leach the remaining organic products. According to the chemical conditions, especially redox level, humic and fulvic acids that may be released by the sediment will vary in proportions with the maturity of the sediment (Orem and Hatcher, 1987).

4.2   Properties of the organic matter

Humic substances, like all biopolymers (proteins, polysaccharides, etc.) show four properties that account for their behaviour (Sposito, 1984):

– Polyfunctionality: the existence of a variety of functional groups and a broad range of functional group reactivity;

– Macromolecular charge: the development of an anionic character on a macromolecular framework, with the resultant effects on functional group reactivity through molecular conformation;

– Hydrophilicity: the tendency to form strong hydrogen bonds with water molecules solvating polar functional groups like COOH and OH;

– Structural liability: the capacity to associate intermolecularly and to change molecular conformation in response to changes in pH, redox condition, electrolyte concentration and functional-group binding.

51

Figure 13. **Evolution of the organic matter during diagenesis (after Tissot and Welte, 1978)**

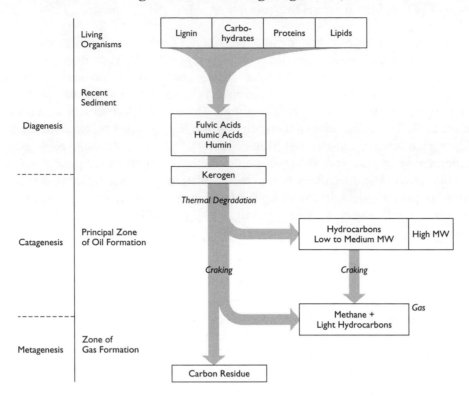

The degree of complexity resulting from those four properties is much larger for humic substances than for other biomolecules, as they reflect the behaviour of interacting polymeric molecules instead of the behaviour of a structurally well-defined single type of molecule.

The soluble organic matter is known to form strong complexes with metal ions. It is held responsible for the modification of toxicity in heavy metals by changing their bioavailability through the formation of either soluble or insoluble complexes (Yong *et al.*, 1992). That property also influences the trace elements of the porewater chemistry.

Organic matter bound to clay particles presents a reactive surface to dissolved solutes in the solution. As a consequence, it may play a role in the buffering of proton and metal cation concentration in the porewater solution. That is achieved via cation exchange, involving proton-bearing surface functional groups and dissolved cations. The most important functional groups in soil humus and their structural formulae are shown in Figure 14.

The association between the solid phases and the insoluble organic matter is not very well understood because the structure of the organic matter is poorly defined. Nevertheless, the same structural groups, as defined in Figure 14, are also responsible for the binding of the organic matter and the clay particle. That may be due to many mechanisms, including cation exchange, protonation, anion exchange, water bridging, cation bridging, ligand exchange, hydrogen bonding and van-der-Waals interactions. Because of the complexity of the subject, a review of the clay/organic-matter interactions is beyond the scope of this report (see *e.g.,* Theng, 1974).

Figure 14. **The organic surface functional groups in soil clays (after Yong *et al.*, 1992)**

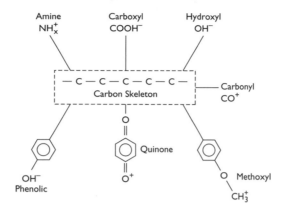

5.   CLAY ENVIRONMENTS RELEVANT TO THIS STUDY

This report aims to review the possible artefact related to the extraction of water and solutes from clay-rich environments with low water content. The targeted applications are a number of sites currently under investigation in different countries. Throughout the report, cases where the examined techniques have been applied to such environments are reported. Table 5 summarises the main characteristics of those study sites in order to provide a quick reference to their properties.

Table 5. **Summary of the main characteristics of the study sites mentioned in this report**

Site name	Stratigraphic unit (name)	Stratigraphic unit (age)	Clay mineralogy (% weight)[4]	CEC (meq/100g)	Water content (% dry weight)	Water type	TDS mg/l	References
Tournemire (France)	Toarcian/ Domerian	Jurassic (Lias)	Clay min. 20-50% (I 15-25, C 2-5, K 5-20, I/S 0-15); Qz 10-25%; Cc 10-52%; Py 0-3%; Org. C 0-6%	8-12 total rock 12-25 < 2 μm fraction	1-4	Na Cl (HCO$_3$)	~ 1,300	NEA/SEDE, 1998; Boisson et al., 1998
Haute Marne/Meuse (France)	Callovo-Oxfordian	Jurassic (Dogger/ Malm)	Clay min. 40-45% (I 0-30, S 10-20, C 5-20, K 20-30, I/S 5-10, C/S 5-10); Qz 25-30%; Cc 25-30%; Py < 1.5%	10-20	4-8	Na Cl (SO$_4$)	>4,000	NEA/SEDE, 1998
Gard (France)	Albian/Aptian	Cretaceous	Clay min. 30-40% (S 30-40); Qz 30-50%; KF traces; Cc 10-25%; Org. C <1%	20-40 < 2 μm fraction	<10	Na Cl (SO$_4$)	> 10,000	NEA/SEDE, 1998
Mont Terri (Switzerland)	Opalinus ton, Opalinus clay	Aalenian	Clay min. 40-80% (I 12-25, C 3-18, K 15-37, I/S 5-20); Qz 10-32%; KF 0-6%; Ab 0-3%; Cc 4-22%; Py 0-3%; Org. C 0-0.5%	10-14.5	3-7.5	NaCl SO$_4$	10,000-20,000	NEA/SEDE, 1998; Pearson et al. (in preparation)
Wellenberg (Switzerland)	Palfris formation	Berriasian/ Valanginian	Clay min. 10-50% (I 3-20, C 2-12, K 0-15, I/S 2-17); Qz 5-21%; Ab 0-2%; Cc 20-80%; Py < 2%; Org. C 0.2-1.4%	8.5	0.5-1.5	NaHCO$_3$ Na Cl	2,000 ~12,000	NAGRA, 1997 NEA/SEDE, 1998
Serrata (Spain)		Pliocene	Clay min. 93% (S 93%); Qz 2%; Pl 3%; Cr 2%; KF traces; Cc traces	110	25-28	Na Cl (SO$_4$)	4,000	ENRESA, 1998; Fernández et al., 1998
Mol (Belgium)	Boom clay	Rupelian	Clay min. 50-60% (I 20-30, S 10-20, C 5-20, K 20-30, I/S 5-10, C/S 5-10); Qz 20-30%; KF 5-10%; Cc 1-5%; Py 1-5%; Org. C 1-5%	30 ± 3.9 (Ag-TU) 24.4 ± 3.4 (Sr) 23.3 ± 3.1 (Ca)	19-24	Na HCO$_3$	1,500	NEA/SEDE, 1998
(United Kingdom)	Oxford Clay	Upper/Middle Jurassic	Clay min. 40-58% (I 14-33, C <5, K 14-33, I/S traces); Qz 11-28%; KF traces; Cc 0-2%; Py 0-3%; Org. C 1-5%		9-20	Na Cl SO$_4$	27,000	NEA/SEDE, 1998

4.   C = chlorite; Cc = carbonates; Cr = cristobalite; C/S = interstratified chlorite/smectite; I = illite; I/S = interstratified illite/smectite; K = kaolinite; KF = K feldspar; Org. C. = organic carbon; Pl = plagioclase; Py = pyrite; Qz = quartz; S = smectite.

PART II

EXPERIMENTAL METHODS

1. FIELD TECHNIQUES FOR FLUID EXTRACTION AND CHARACTERISATION

1.1 Piezometer and borehole drilling

In-situ water-extraction techniques normally involve drilling. A good review of the available drilling tools for unconsolidated and consolidated sediments is reported in Wijland *et al.* (1991). Drilling operations should be carefully planned in order to avoid a permanent contamination of the environment from drilling fluids (Tshibangu *et al.*, 1996). In fact, in very low water-content systems, long delays are required before purging the system from that type of contamination, and cases are reported where a representative formation fluid could never be recovered (NAGRA, 1997). In principle, the use of an "equilibrated" fluid would be recommended, to which a tracer may be added. In practice, solutions are selected based on their availability and normally consist of surface waters. The water flowing from the Cernon fault, a major discontinuity intersecting the formation, has been used in most drilling work in the Tournemire Tunnel. That water naturally contains tritium, and that component is consequently considered the reference tracer. It is worth noticing that, in the absence of intersecting fracture systems, no water flowing from boreholes could even be collected in the Tournemire claystone. All investigations are conducted using core samples. Contamination of those samples is restricted to the outer rim, and may easily be eliminated.

An alternative to the use of drilling fluid is air drilling. That solution is considered in an increasing number of studies (Yang *et al.*, 1990; Griffault *et al.*, 1996; Thury and Bossart, in preparation), at least for relatively short boreholes. It has the advantage of being far less contaminating and allowing to see actually humid zones intersected by drilling. On the other hand, drilling has to proceed slowly and costs are considerably increased. The main disturbance consists of a desaturation of the environment, and consequently, in the increase of oxidation processes. Besides, the lack of added tracers does not allow understanding of the entity and of the duration of the induced disturbance. A small contamination from hydrocarbons, possibly coming from the compressor oil used for air drilling, has been detected in the organic matter extracted from the clay samples of the Mont Terri Tunnel (Thury and Bossart, in preparation). On the other hand, no contamination of the water collected in the boreholes or squeezed from the rock samples has been observed.

One of the main objectives of the ARCHIMEDE project at the HADES facility (Mol, Belgium), was to obtain mechanically, thermally and physico-chemically undisturbed clay samples. Some piezometers have been introduced by pushing a set of stainless-steel cutting bits into the clay formation at elevated pressure (20 MPa). That technique, although only applicable in "soft clays", allows the recovery of clay samples without torsion or rotational movements causing fracturation and local heating of the clay. The main disturbance is the high compression of the clay at the cutting edge.

In order to eliminate the strongly compressed clay debris, the borehole had to be rebored after each sample collection, thus partially limiting the core recovery.

1.2   Piezometers and borehole equipment

Boreholes to be used in investigations in low-permeability media may be divided into two groups, based on the technical requirements for different depths (Wijland *et al.*, 1991). Shallow boreholes, up to 600 m deep, may be drilled without the use of casing, with the exception of the stove-pipe, to be placed in the upper part in order to prevent flushing of the superficial sediments. The use of PVC tubing, chemically inert with respect to the solutions is suggested. Minimum diameters for observation wells are 1 inch, while 2-inch diameters at least are required to perform production tests. Deeper boreholes have to be protected by steel casing and require additional backfill material (gravel or cement) to fill the space between the rock wall and the casing. If steel and concrete are used, that will cause an important contamination of water samples. If a stainless-steel well casing is used, the contamination is reduced and mainly concerns trace elements, but costs are increased.

Most of the *in-situ* migration studies performed at the HADES facility (Mol, Belgium) were performed using piezometers. That allowed the SCK staff to acquire an impressive expertise in that field. Geochemical investigations through piezometers were selected mainly because early attempts to obtain porewater from the clays via ultracentrifugation or squeezing had failed, despite the rather high water content of the formation (~ 20% weight) (Beaufays *et al.*, 1994). In addition to conventional piezometers using a backfill material, self-sealing piezometers have been installed, relying on the natural convergence of the Boom clay. The main advantages of those piezometers consist in the absence of filling material, the possibility of drilling in all directions, minimising contamination and a small dead volume allowing a relatively fast water renewal in the piezometer. More than 100 piezometers of different types (single-screen, multi-screen, long cylindrical screen, flat-screen) were installed, in addition to piezometer-nests (Figure 15) with 19 filters each. Piezonests are based on the same self-sealing principle and allow to determine the porewater pressure as a function of the distance from the gallery and to run simple migration experiments. Besides, they allow sampling of porewater for chemical and isotopic analyses.

The piezometers are used to measure the porewater pressure and flow rate in order to determine the hydraulic conductivity. The values found for the different set-ups are in good agreement, despite the large difference in geometry and size of the filters (up to four orders of magnitude). In addition, *in-situ* values for hydraulic conductivity are the same as those obtained on small-scale laboratory experiments, in contrast with what is reported in current literature (Beaufays *et al.*, 1994).

Samples collected in the self-sealing boreholes are considered fairly representative of the porewater chemistry, allowing the description and modelling of the solid/water interaction (Griffault *et al.*, 1996). Stable isotopes instead appear to be affected by a fractionation very similar to that observed for water samples obtained from the clays by sample squeezing. That similarity is not surprising if we consider that, because of the hydraulic and lithostatic pressure, the plastic clay around the piezometer actually squeezes the waters into the collecting cells. The data points plot below the global meteoric water line.

In the frame of the ARCHIMEDE project funded by the European Community, a big effort has been put in the installation of seven especially designed boreholes, among which two sterile piezometers for porewater sampling and a 20-m borehole for clay sampling in aseptic and anaerobic conditions for microbiological investigation.

Figure 15. **Schematic design of a piezometer nest at the HADES facility (Mol, Belgium)**
(after Beaufays *et al.*, 1994)

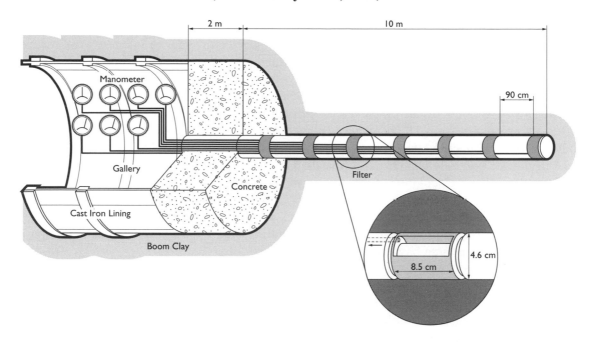

The design of sterile borehole drilling required a very careful examination of the sterilising procedure for drilling and sampling tools (Griffault *et al.*, 1996). When possible, tools were sterilised with a 1.5% formaldehyde solution. The borehole for clay collection was drilled using sterilised cutting bits to preserve the sample from external microbial contamination. Piezometers for porewater sampling were air-drilled. Air represented a possible source of contamination, since it could not be filtered. Tests were conducted to check the degree of air penetration in the cores. Those showed little contamination but, for more precaution, the outside rim of the samples was trimmed before packing. Piezometers were preserved with previously sterilised castings, and included five filters each. The seals of the castings were welded with a blowpipe. In one of the piezometers, the tubing was heated at approximately 180°C before emplacement, thus sterilising locally the external surface of the borehole. That sterilising method drastically increased the pyrite oxidation, due to the elevated temperature and the presence of oxygen during the piezometer conditioning. Five years later, the effects of the chemical disturbance were still detectable (H. Pitsch, CEA, personal communication).

1.3 Field techniques for fluid extraction

1.3.1 Under-vacuum water extraction

Under-vacuum samplers were mainly developed for the collection of soil solutions. We will not describe those devices in detail, as their design cannot be adapted directly to the relevant clay environment. Nevertheless, some global considerations on the disturbance induced by those techniques may also apply to our case studies. Litaor (1988) reviewed all the available types of soil-water collectors. The problem of the disturbance of the medium is considered, since the water movement may be easily disturbed in soil by the preferential path created during the installation and by the modified suction gradient. The main conclusion of his review concerning under-vacuum ceramic cups is that the applied suction should not exceed the soil suction, in order to minimise the loss of volatile compounds. Important changes are observed in pH and redox potentials of the collected solutions, and eventually a precipitation of carbonates within the cups may occur. Besides, ceramics seems to influence by retention or release a number of chemical constituents (nitrates, phosphates, potassium, aluminium and calcium). Other materials (porous Teflon, fritted glass) are also considered.

The principle of slightly underpressurising the device in order to increase fluid recovery in low-permeability environments has been applied to many installations. That aims also to reduce the effects of oxygen on the collected solution. In highly-indurated rocks, the borehole end may be isolated with a packer and slightly underpressurised. That solution has been adopted for example, in the Mont Terri Tunnel where boreholes are drilled almost vertically and their end is isolated with a packer (Thury and Bossart, in preparation). The borehole is purged with nitrogen and slightly underpressurised. The draining water is collected after several weeks, using an inert type of tubing connected to a funnel located inside the borehole chamber. The installation also allows the gas sampling in the headspace, possibly enabling the recalculation of the modifications of the water chemistry due to degassing.

Within the framework of the investigations on the Tournemire site, B.A.-Tortensson (BAT) sampling devices (Torstensson, 1984; Torstensson and Petsonk, 1985) were installed (Boisson *et al.*, 1998). Those consist in three basic components:

–   A permanently installed sealed filter tip attached to an extension pipe (*A*). Filter tips are of different shapes and materials depending on their use (Figure 16). The body may be in thermoplastic, stainless steel or brass, while the filter is made of high-density polyethylene, sintered ceramic or porous polytetrafluorethylene (PTFE).

–   A disposable double-ended hypodermic needle (*B*).

–   A pre-sterilised evacuated sample vial of glass (*C*).

Both the filter tip and the vial are sealed with a pre-stressed disk of resilient material, *e.g.*, synthetic rubber. The sampling vial and the needle are mounted on a sampling probe that is lowered in the pipe. Here the probe connects to the cap of the filtering tip, the double-ended needle providing a temporary leak-proof connection to the sampling vial. Due to the vacuum in the vial, both water and dissolved gases may be sampled in the *in-situ* conditions. Filter tips may be installed in boreholes, and may be placed under vacuum. The same filtering tip may be used for measuring pore pressure and tension. In that case, a pressure transducer is connected to the water-filled chamber via a needle. BAT devices are mainly designed for soils and soft rocks. In the case of Tournemire, the installed devices proved unable to recover water samples, because depth and pressures were too large (Boisson *et al.*, 1998).

1.3.2   *Under vacuum gas extraction*

In-situ devices for soil-moisture extraction have been developed, based on the assumption that, in the unsaturated zone, water vapour and liquid should be in equilibrium. Thoma *et al.* (1978) described a soil-air suction probe, consisting of a stainless-steel tip that is hammered down to the selected depth in the soil. The tip is connected to a pumping system (flow rate ~ 30 l/min), that pumps the moisture-saturated soil air through a molecular sieve (0.4 nm). The soil moisture is later re-extracted quantitatively in laboratory by heating the molecular sieve to 400°C in a vacuum system and by freezing the released water in a dry-ice cold trap. That device has been first applied for tritium peak detection and deuterium analysis in a sand dune. The resolution of that method, compared to the auger-sampling method, was surprisingly high (a few centimetres), considering that 300 l of soil air had to be pumped in order to obtain enough water for the analysis. Assuming a porosity of 0.25, that would correspond to

Figure 16. **BAT Points (after Torstensson and Petsonk, 1985)**

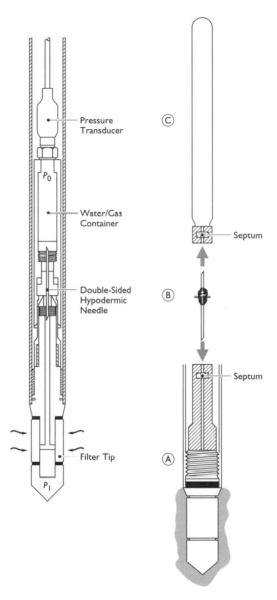

the amount of air contained in a 1-m radius spherical volume of soil around the suction tip. The authors explained the phenomenon by an extremely rapid water/vapour molecular exchange occurring in the proximity of the tip. That assumption was also confirmed by laboratory experiments. Tests using a very similar probe on tritium-tagged plots were reported by Saxena and Dressie (1983). Comparison between soil-air suction and coring data showed an agreement in the distribution along the profile, but lower concentrations were obtained with the former technique. That is attributed to field variability and/or dilution of the tritium activity by atmospheric moisture leaking during the sampling. Finally, Allison *et al.* (1987) conducted a very careful field and laboratory study on both oxygen-18 and deuterium, comparing water extracted by soil coring and vacuum distillation, with water vapour collected with a probe. Additionally carbon dioxide of the soil gas was sampled and analysed, as another tool for obtaining the oxygen-18 value of soil water (see Chapter II § 3.5.1). They showed that, although profiles displayed the same trends, isotopic fractionation factors would vary considerably. Laboratory data confirmed that the equilibrium assumption was verified and that the fractionation factor did not depend on the soil-matrix suction. They demonstrated that discrepancies were due to imperfect sampling techniques, but that those artefacts may not easily be overcome. They reported those results "because they show sufficient agreement to warrant further study".

As an application to saturated clay environments (Tournemire formation), Moreau-Le Golvan (1997) reported an attempt of *in-situ* carbon dioxide extraction for carbon-14 analysis. Two freshly air-drilled boreholes were equipped with a double-packer system (Figure 17). The first chamber was filled with an overpressure of nitrogen to prevent leakage from the atmosphere in the sampling cavity. Pumped carbon dioxide was trapped using liquid nitrogen, both in static and dynamic vacuums. Prior to the installation of packers, the hole was flushed with argon and pumping was performed for 1 h before collecting samples to eliminate possible contamination from air-drilling. Despite those precautions, results showed a low, but detectable carbon-14 content in all samples. A high fractionation of carbon dioxide was obviously observed, as proven by the variation in stable isotopes.

Figure 17. **Schematic drawing of the *in situ* gas extraction experience (after Moreau-Le Golvan, 1997)**

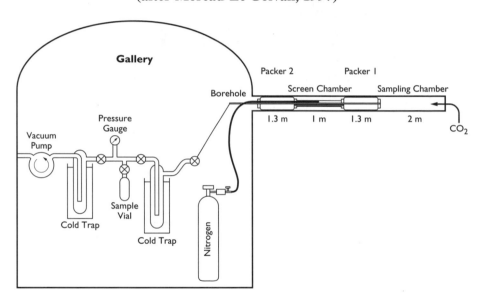

1.3.3   Dialysis cells

In-situ equilibrium dialysis samplers, also called "peepers", were first described by Hesslein (1976) and are now widely used for porewater sampling in sediments. They have replaced the original idea of directly using dialysis bags (Mayer, 1976) that proved too long to equilibrate. They rely on the equilibration established between de-oxygenated, de-ionised water contained in the sampler and the sediment porewater through a dialysis membrane. Teasdale *et al.* (1995) reviewed their design, the materials and membranes used, and the physical process affecting the equilibration. Sampler design has to be a compromise between several requirements, including the sample volume, the time required for reaching equilibrium and the need to cause minimal disturbance when inserted in the sediment. Normally, volumes range between a few millilitres and 30 ml. Application to sampling anoxic porewaters for trace-element analysis is reported in a number of papers (see *e.g.,* Apte *et al.*, 1989; Steinmann and Shotyk, 1996; Azcue *et al.*, 1996; Urban *et al.*, 1997). The low sample volumes combined with the low concentrations of many metals makes that analysis extremely challenging.

The use of thin films (Davison *et al.*, 1991) rely on the same diffusive equilibration principle. In that case, rather than confining the solution to a compartment, a thin-film or gel is used to provide the medium for solution equilibration. That has the advantage of much faster equilibration times and higher resolution, but needs to be analysed in its solid form by proton-induced X-ray emission (PIXE).

Dialysis membranes may readily prevent oxidation of the water sample when isolated from its medium. The rate of oxygen diffusion from the atmosphere into the chambers is quite high (4 µmol/minute, in the referred experimental conditions; see Carignan, 1984), and that may result in iron-hydroxide precipitation. The chambers must therefore be subsampled as soon as possible (within 20 min).

Although reported studies focus on the application to high water-content systems, an attempt to transpose that technique to clay-rich environments has been performed in the framework of the ARCHIMEDE project in Mol. The purpose of the experiment was to install a device completely metal-free for heavy-metal analysis in collected samples. An especially designed borehole (Figure 18) was initially conceived in reinforced and vulcanised rubber. That design did not prove resistant enough to the pressure, and the tube was squeezed soon after its emplacement. A second borehole in more rigid, pressure-resistant material (such as PVC) was installed. A screen in porous ceramics was adopted in order to minimise the risk of collapse and the possibility of clay intrusion in the dialysis chamber, while maintaining its inert characteristics. The dialysis cell was placed into that filter chamber, located at the end of a 5-m long nylon tubing. The system was isolated with inflatable packers in order to allow a partial restoration of the *in-situ* conditions. The main problem was the high-pressure difference between the fluid in the screen chamber and the equilibrating fluid in the dialysis cell. The packer would automatically open or close in order to regulate the pressure inside the borehole preventing the membrane from tearing. Unfortunately, only one water sample could be obtained, and analysis could not completely validate the installation for trace metals. Fractionation is thought to occur when water passes through the pores of the ceramic filter, before crossing the dialysis membrane. On the other hand, the use of dialysis membranes directly in contact with the clay is not possible because of their low mechanical strength.

1.4   In-situ physico-chemical measurements

Ion-selective electrodes have been extensively studied for the determination of the fluid composition in high water-content systems (see Frant, 1997 for a review). Recent developments of microelectrodes (De Wit, 1995; Kappes *et al.*, 1997) have not yet found an application in low water-content systems. They are rather used for establishing "*in-situ*" profiles in sediments (Hales *et al.*, 1994; Hales and Emerson, 1996), or for measuring pH and other parameters on very small quantities of extracted water (*e.g.,* water extracted from squeezing cells).

1.4.1   pH measurements

Among the numerous physical and chemical parameters to be considered, pH is of particular interest for the chemistry regulation of pore fluids since it appears in most element-speciation equilibria. Sampling techniques with postponed analysis raise the problem of pollution risks. Possible modifications of the characteristics by induced disturbances are particularly relevant for fluids that are poorly buffered when they are no longer in contact with the mineral assemblage of the formation. In the case of clay formations, low water content and high interstitial pressure are additional constraints. As a consequence, *in-situ* measurements have been designed using different techniques.

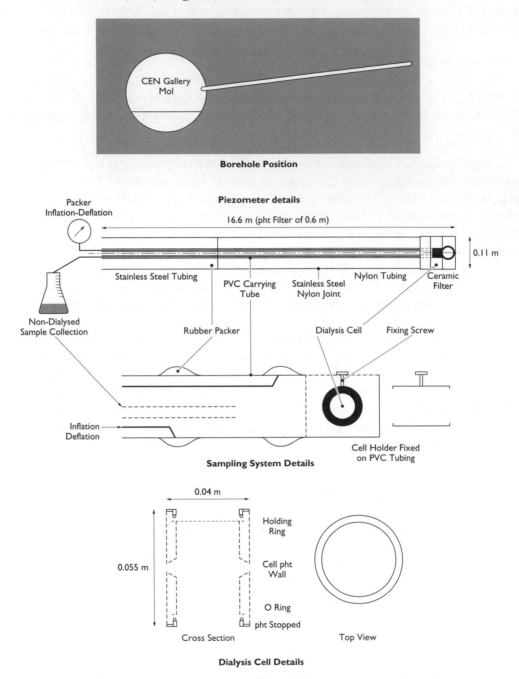

Figure 18. **Installation of the dialysis cell borehole at the HADES facility (Mol, Belgium) (after Griffault *et al.*, 1996)**

Early attempts to use glass electrodes directly in the piezometers failed because of pressure problems (glass electrodes are very fragile) and calibration problems (Griffault *et al.*, 1996). In fact, the impossibility of extracting the electrode from the borehole to check calibration without heavily disturbing the system, combined with the long amount of time required for pH equilibration, produced unreliable results.

Installing the electrode outside the borehole is another option, provided that a circulation pump homogenises the fluid. In that case, the electrode may be easily isolated from the circuit for calibration without inducing major disturbances of the system. In the framework of the Mont Terri project, that solution was adopted (Thury and Bossart, in preparation). A dedicated borehole was filled with a synthetic solution close to equilibrium with the rock (see Chapter III § 3). pH is continuously monitored on the circulating fluid until it reaches a constant value.

pH measurement within the borehole may be obtained using a fibre-optic pH sensor (Motellier *et al.*, 1995). That approach has been especially developed by the French Atomic Energy Commission (*Commissariat à l'Énergie Atomique – CEA*). pH is measured via the absorbance of a light signal sent by an optical fibre (Figure 19). The sensing tip (optode) is composed of a semi-bead of resin impregnated with phenol red and immobilised at the distal end of a 1 mm per 20-m optical fibre (Motellier *et al.*, 1993). The optode is connected to an emitter-receiver apparatus (Optolec H). That consists in three modulated light-emitting diodes (LED) of monochromatic light that are set to avoid any interference with the surrounding light. λ_1 is set at the wavelength of the isosbestic point. The signal recorded corresponds to the reference for the quantity of indicator located at the end of the optode and accounts for the degradation or the unsteady binding of the indicator. λ_2 is the measuring wavelength and is set at the maximum absorptivity of the alkaline form of the indicator. λ_3 is chosen close to the infrared in a region where the indicator is non-absorbent, whatever the pH value might be. That signal allows for the correction of the noise or drift of the optical line, the change in the position of the bead or the torsion of optical fibres.

Figure 19. **Fibre-optic pH sensor (after Griffault *et al.*, 1996)**

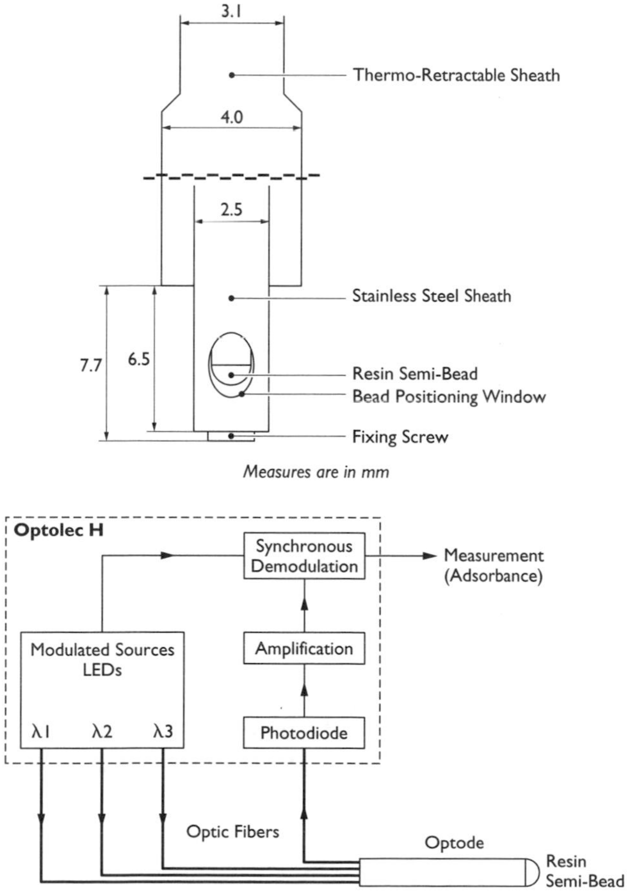

That system has been tested against other pH-measurement techniques (batch and in-flow with glass electrode) at the HADES facility (Pitsch et al., 1995b). The interstitial fluid is non-buffered with respect to acidity and is subject to degassing when sampled. The optode was introduced in the borehole via a stainless steel guideline in the chamber of a 15-m long piezometer equipped with a stainless-steel filter. The guideline was tightly closed in order to allow the pressure restoration to its nominal value of 17 bars. The data acquisition lasted about 200 h, with one acquisition per hour. The pH value (8.21), compared with that obtained using an in-flow glass electrode (8.19), showed a very good agreement. In addition to its pressure resistance, the device had a good response even if the water around the sensing tip was not renewed. It may consequently be used also in situations where discharge is lower than 1 l/h (Pitsch et al., 1995b).

1.4.2   Eh measurements

Attempts to obtain an *in-situ* Eh measure have been performed, but results seem to be mainly restricted to fluids that display a measurable equilibrium potential imposed by a dissolved redox couple (buffered solutions) or a stable mixed potential imposed by two different redox couples.

In the framework of the CERBERUS project (Boom clay), the water flowing from two piezometers installed close to the radiation source was analysed *in-situ* for its pH and Eh (Noynaert et al., 1997). Both values were measured in a cell kept at 80°C, the *in-situ* temperature of the water, to avoid gypsum precipitation. Results were compared to those obtained *in-situ* at 22°C and those measured in the lab at 22°C. Eh values were also calculated from the Nernst law[5], knowing the pH and H_2 concentrations that were independently measured. The comparison shows that Eh calculated values are systematically lower than the measured ones. That may reflect the fact that the hydrogen produced during the experiment is not the species controlling the Eh. In addition, the high-temperature Eh values are higher than the low-temperature ones, that behaviour being in contradiction with the Nernst equation. That is explained by the intervention of other electroactive species, such as thiosulphate, free sulphide and Fe^{2+}/Fe^{3+}, arising from pyrite oxidation.

In the absence of electroactive species, the measured potential is unstable since very small disturbances (concentration fluctuations, electrode surface modifications) cause the potential to change abruptly from one value to the other (Pitsch et al., 1995a). A flow-cell technique tested on samples taken from different clay/water environments proved to be more reliable than batch measurements, even in the case of reducing environments, provided that the anaerobic conditions are respected. More reliable Eh values may be obtained by modelling and/or extrapolating the data obtained for the redox couples that are known to be present in the fluid, even if they are not in a sufficient quantity to provide a stable electrode potential ($< \sim 5 \mu M$) (Beaucaire et al., 1998).

In the near future, a non-metallic piezometer for Eh and trace-element measurements will be installed in the underground laboratory in Mol (R. Gens, ONDRAF, personal communication).

1.4.3   Temperature

The use of an optical fibre for borehole-temperature logging is reported by Förster et al. (1997). Although resolution is 5 to 10 times lower than conventional techniques, that system quickly

---

5. $E = E_0 - 2.3(RT / nF) \log P_{H_2} - 2.3(RT / 2nF) \text{ pH}$
        $E = -0.030 \log P_{H_2} - 0.059 \text{pH} (25° C)$

responds and is not affected by problems related to variations in cable resistance, disturbing electric currents and improper isolation.

1.5   Field techniques for indirect fluid characterisation

1.5.1   Wireline logging

Borehole-logging tools represent a mature technology that is widely used in oil and gas exploration (Schlumberger, 1997). The application of those techniques to the assessment of radioactive-waste repositories is currently being investigated. A combination of wireline logging, seismic, hydrologic and geomechanical testing techniques may provide valuable information for site characterisation. That includes fracture and fault detection and mapping, the physical properties of the rock (lithology, stratigraphy, porosity), geochemistry (rock-forming elements), hydrologic properties (conductivity, transmissivity), *in-situ* stress and geomechanical properties.

Since the 1960s, Schlumberger and other service companies (Kenyon *et al.*, 1995) have been applying nuclear magnetic resonance (NMR) to the *in-situ* determination of rock porosity, moisture content and amount of free and bound water. That technique is currently undergoing testing on low-permeability clay formations at the ONDRAF and NAGRA sites (Win *et al.*, 1998; Strobel *et al.*, 1998).

NMR has been widely used to determine the properties of adsorbed water on clays (Sposito and Prost, 1982). The hydrogen atom has a magnetic moment that may be oriented by a strong magnetic field. The combinable magnetic resonance (CMR™[6]) tool developed by Schlumberger is constituted by a powerful permanent magnet that polarises the non-lattice bound hydrogen atoms within the pore fluids of the formation. In addition, a directional antenna emits tuned radio pulses at right angles to the main polarisation field. When the pulsed radio signal is turned on, the free hydrogen atoms align with a magnetic field transverse to the polarising field of the permanent magnets. When the pulsed radio signal is turned off, the hydrogen atoms attempt to realign with the polarising field of the permanent magnets. Those atoms precess as they try to return to their original orientation, emitting a radio signal that is measured by the CMR tool. The amplitude of the signal is proportional to the number of hydrogen nuclei. The associated decay rate, known as "transverse relaxation time" (T_2) is related to three main mechanisms: collisions of the protons with the pore surfaces, molecular diffusion in magnetic-field gradients and bulk-fluid relaxation. Since the two last mechanisms may usually be neglected or taken into account in any correction calculations, the amplitude of the signal is proportional to porosity, assuming that all the pores are saturated with water, and the overall decay rate reflects pore-size distribution.

In Figure 20, the recorded signal amplitude may be separated in a time-series-dependent T_2 values by a mathematical inversion process, producing the T_2 distribution. The curve shape represents the pore-size distribution, while the area under the curve represents the CMR porosity. Laboratory experiments on core samples may help in selecting a T_2 cut-off delineating water bound to the solid phases from free water. Typical cut-off values for clastic rocks are around 33 msec for T_2.

The CMR tool is designed for borehole logging. Its position is constrained to the side of the borehole by an excentralising spring permitting investigation of the formation that should be less disturbed by drilling effects. Data may be compared with hydraulic conductivities, grain-size distributions and other parameters obtained both on core and on borehole logging (Figure 21).

---

6.   Trademark of Schlumberger.

Figure 20. **CMR signal amplitude processed to provide T$_2$ distribution and its interpretation in terms of porosity and types of water (modified after Kenyon *et al.,* 1995)**

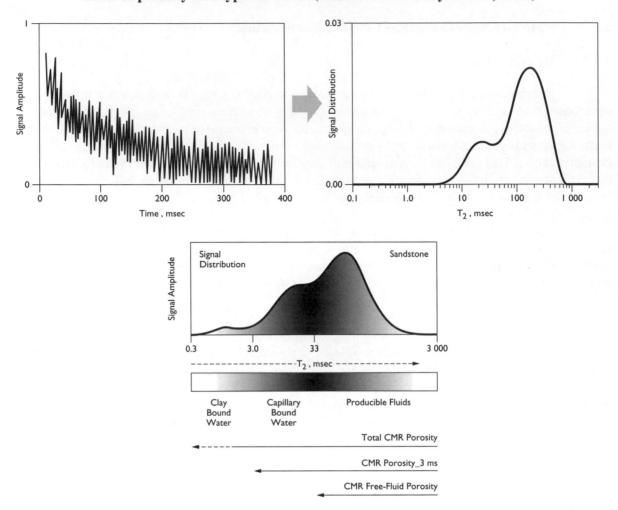

The *in-situ* NMR technique has been successfully applied to the oil industry and seems to give very promising results, even for fine-grained sediments. Uncertainties are related to the estimation of T_2-derived free fluid line and the estimation of constants necessary to calculate permeability, both of which are based on semi-empirical test results established for sandstones. An accurate calibration of CMR data to core permeabilities, together with grain-size analysis, may improve the tool ability to log those parameters on a continuous basis. Finally, the integration of CMR with other logging tools, such as the accelerator porosity tool (APT™), an epithermal neutron tool used to estimate total porosity, the elemental capture spectrometer (ECS™), a geochemical logging tool that calculates elemental concentrations for silicon, calcium, iron, sulphur, titanium and gadolinium, and other tools (Croussard *et al.,* 1998) should successfully quantify hydrologic properties and detect heterogeneities in argillaceous formations. It may be considered as an alternative or complement to additional coring.

1.5.2   *Monitoring of the drilling fluids*

Monitoring of the drilling fluids is a technique that has been developed for crystalline rocks, mainly in order to identify water-bearing fractures (Vuataz, 1987; Vuataz *et al.,* 1990). Its application to sedimentary rocks was reported by Aquilina *et al.* (1993a; 1993b; 1994; 1995; 1996) as a tool for the

Figure 21. **CMR permeabilities shown with Mol-1 well-core hydraulic conductivities, grain sizes, CMR and APT porosity curves, T_2 distribution and hydrogeologic units (after Strobel *et al.*, 1998)**

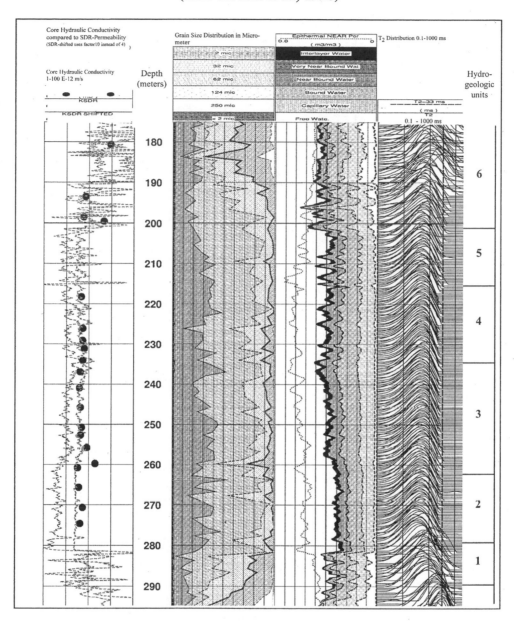

determination of porewater composition. Those studies concern the Balazuc borehole, 1,725 m deep, intersecting low-porosity rocks with no water inflow and the Morte-Mérie borehole, 920 m deep, reaching the Permian basement. The WELCOM (well chemical on-line monitoring) system (Figure 22) consists in chemically logging the concentrations of the major and trace elements in the drilling fluid. Fluids, continuously pumped and filtered through a tangential filtering device, both at the inflow and the outflow of the well, are analysed in a field laboratory. After processing and correcting the signal, the difference between the inflow and outflow is attributed to the influence of the crushed formation at the bottom of the borehole. Chemical inputs from the formation may originate from mineral dissolution and/or from interstitial solutions. When the origin of input may be attributed to porewaters, the signal may be used to calculate their composition.

Figure 22. Schematic layout of the continuous chemical logging operation (after Aquilina *et al.*, 1996)

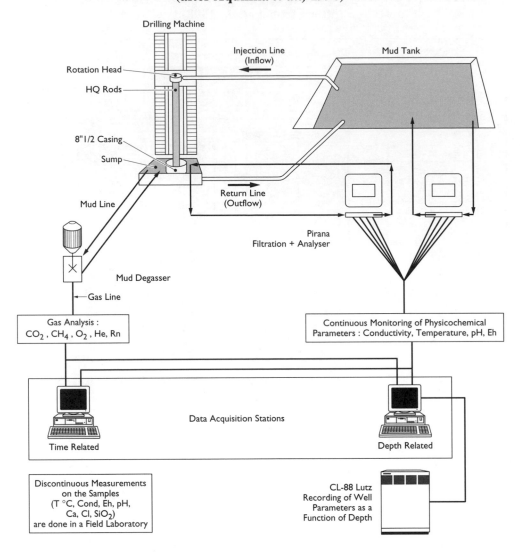

Monitoring of the drilling fluids may be compared to the leaching of the drilled rock column (see Chapter II § 3.3). However, parameters differ substantially from on-core leaching, as dilutions are made with surface water containing bentonite, polymers and sodium hydroxide instead of distilled water. In addition, the interaction time is shorter and occurs at an uncertain solid/liquid ratio. Data obtained by on-core laboratory leaching with deionised water are compared with WELCOM chemical inputs. Despite the number of assumptions and corrections to apply, authors reported a fairly good agreement between the two methods, and even a higher sensitivity for the latter (Aquilina *et al.*, 1995; 1996). Uncertainties arise from the difficulty to control the volume of rock leached by drilling and the fluid/clay interactions. Since pore fluids show much contrasted compositions and the internal consistency of the data is assured, the authors concluded that results might be used for qualitative interpretation. In our opinion, the type of drilling fluid used in that study is extremely aggressive and disturbing, and should be avoided if geochemical considerations have to be made. Its use may also disturb solid samples, thus preventing their use for geochemical analysis. That system is not adapted to very low water-content systems.

2. ROCK SAMPLING, STORAGE AND PRESERVATION

2.1 Evidence of artefacts

The Eh and pH conditions existing in any deep rock formation have a significant influence on almost all the critical parameters determining the behaviour of the system with respect to radionuclide migration. The sampling process, isolating portions of rock or fluid from its environment, may induce changes in the sample characteristics that may be virtually impossible to estimate with any certainty. Closely associated with that problem are the issues of long-term storage, handling and preparation of samples.

A large amount of the literature has dealt with the changes in porewater compositions of soils and marine sediments. Effects of changing temperature, pH (mainly due to degassing) and Eh are reported with a view to evaluating their importance and consequences (Hulbert and Brindle, 1975; Lyons *et al.*, 1979; Ross and Barlett, 1990; de Lange *et al.*, 1992; You *et al.*, 1996). In low water-content systems, artefacts seem to proceed much slower, due to the lower diffusion coefficients of water and gases in the small porosity.

Other effects, such as dehydration and oxidation, should be prevented with some simple precautions. The main objective is to isolate as soon as possible the sample from the atmosphere. That operation should consequently be performed on site, soon after core retrieval. Air-tight aluminium or stainless-steel cans are commonly used in soil sampling (see *e.g.,* Saxena and Dressie, 1983; Allison *et al.*, 1987). The application to indurated clays is not straightforward, mainly because the hardness of the rock does not allow to fill the container properly and some air may be trapped inside. That is a minor problem in the case of soils because the unsaturated zone is an oxidising environment.

Sample conditioning may be performed by wrapping the rock in aluminium foil and beeswax (Brightman *et al.*, 1985; Yang *et al.*, 1990) or by coating it directly with paraffin (see *e.g.,* Ricard, 1993 and Moreau-Le Golvan, 1997). Another increasingly used solution involves aluminium-plastic bags or foils that may be thermally welded. The sample is introduced in the bag, flushed with nitrogen or mixtures of inert gases, then evacuated and sealed. That type of conditioning seems adequate to preserve of samples for mineralogical and chemical analyses. Despite that, the danger of evaporation is not excluded. In fact, if the atmosphere around the sample is dehydrated, that will create a gradient driving the moisture out. Wrapping the rock in aluminium is highly recommended if the organic matter has to be analysed. Polyethylene bags were found highly contaminating by phtalate release (Griffault *et al.*, 1996).

If the sample needs pre-treatment prior to analysis (crushing, sieving, etc.), those operations should be conducted by minimising both the contact time with the atmosphere and the use of potentially contaminated tools or devices. In order to eliminate the possibility of a contamination from the drilling fluid, it is recommended to trim the outside rim of the cores and discard it. Edmunds and Bath (1976) used a lithium-chloride tracer added to the drilling mud to investigate the extent of fluid invasion into the cores. They showed that for chalk the invasion was limited to the outer 1-1.5 cm, while, in the case of a highly permeable sandstone, samples still showed more than 2% fluid replacement after the outer 2 cm had been removed. They concluded that air-flush drilling methods were more suitable for permeable material investigations. Moreau-Le Golvan (1997) conducted vacuum distillations on Tournemire-claystone samples that had been sealed with wax after recovery and found no change in the isotopic composition between the outer and inner parts of the core, but for more precaution he still discarded the outer 0.5 cm of each sample prior to crushing.

Yiechiely *et al.* (1993) found that the main factors controlling the moisture content of soils collected close to the Dead Sea (Israel) were the percentage of fine-grained material and the salt concentration. The former would limit the evaporative loss of the original solution and enhance water adsorption from the atmosphere. Laboratory experiments conducted on oven dried sediments showed the importance of the salt content in air moisture re-adsorption. The good correspondence between the moisture fluctuations of the laboratory air and the sample is shown in Figure 23. The same samples leached with deionised water to lower their salt content show a much lower uptake of air moisture and almost no fluctuations.

Figure 23. **Air moisture uptake after oven-dried Dead Sea sediments.**
Sample 1 = 143 mg Cl/g of sediment; sample 2 = 102 mg/g;
sample 1a and 2a are the same, leached down to 1 mg Cl/g of sediment
(after Yiechieli *et al.*, 1993)

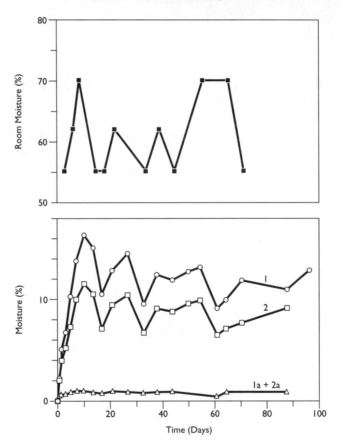

Moreau-Le Golvan (1997) also showed an isotopic effect related to the contact time with the atmosphere, during and after crushing, for the Tournemire-claystone samples. For time intervals ranging from 15 to 60 min, an enrichment in both isotopes is observed. Results, in the deuterium *versus* oxygen-18 plot, fall on a line displaying the typical slope of evaporation lines (Figure 24). Unfortunately, no indications are given on the laboratory air-moisture content, to help evaluate if evaporation prevails over air-moisture uptake for clay-rich materials.

Figure 24. **Influence of the contact time with the atmosphere**
(after Moreau-Le Golvan *et al.*, 1997)

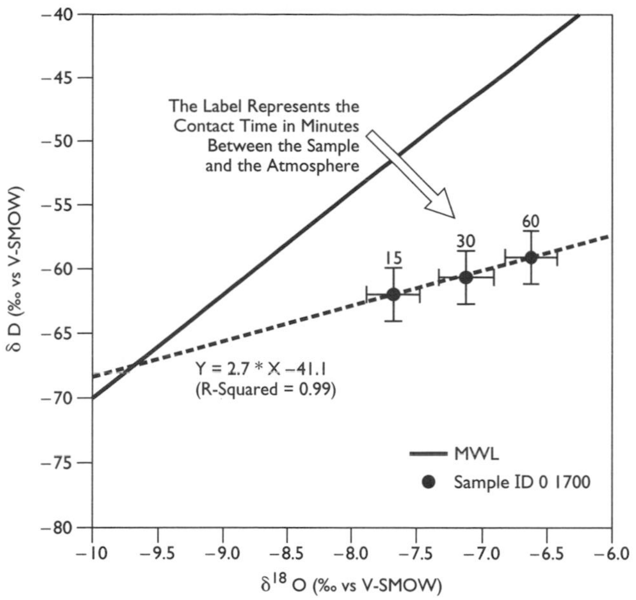

2.2   In-situ freezing and coring

In the framework of the ARCHIMEDE project (Griffault *et al.*, 1996), an *in-situ* freezing test has been conducted. The aim of the experiment was to define possible hydrochemical changes in the case where freezing before the repository excavation would be considered. In addition, alternative techniques of water extraction from the cores have been attempted (see Chapter II § 3.6).

One cubic metre of clay was frozen using six 3-m long freezing elements inserted in the rock in a hexagonal arrangement. The rock prism was frozen at −10°C during two weeks, after which two cores were taken from the central part of the frozen formation. The first core, 48 mm in diameter, was particularly difficult to obtain and a 50-cm sample cut in four pieces was recovered. The second core, 100 mm in diameter, provided two samples from 0 to 0.75 m and from 0.75 to 2 m inside the formation.

Partial defreezing of the samples cannot be excluded because of the temperature increase at the contact point between the drilling head and the formation. Nevertheless, it seems that its extent was very limited. Samples obtained from cores were kept in dry ice before shipping to the laboratory, where they were stored in a freezer at −18°C. Subsequent analysis of water contents and stable-isotope compositions indicate that ice sublimation may have occurred. Unfortunately, no indications are given on the storing conditions inside the freezer and the storage time prior to analysis.

2.3   Noble-gas sampling procedure

For noble-gas analysis, Osenbrück (1996) has designed a new sampling and preservation method (see also Osenbrück and Sonntag, 1995). A piece of rock is taken from a freshly drilled core as quickly as possible. The outer parts of the sample are trimmed over 2 cm in order to eliminate a possible contamination due to the drilling fluid and the degassing effects during core lifting and handling. Samples are placed in a vacuum-tight brass container, then evacuated with a rotary pump and flushed twice in between with pure nitrogen. After four weeks, the noble gases are completely extracted from the porewater into the vacuum by molecular diffusion. They are directly transferred to the mass spectrometer for analysis, the accuracy being about 5-10% for all of them.

3. ON-SAMPLE LABORATORY TECHNIQUES

In the following section, we will examine all the techniques that have been used for extracting water from soil and rock samples. The choice has been made to describe the different techniques, regardless of their application to specific chemical or isotopic analyses. Table 6 lists the techniques under review. Indications on the suitability of a given technique for chemical and isotopic analyses are given, based on the existence of published studies in the field. It is also stated when the use of a given technique is impossible based on the physical principle used.

Table 6. **Use of different water extraction techniques for chemical and isotopic analyses**

Technique	Specifications	Chemical analysis	Isotopic analysis
Centrifugation	Low/high speed	Major and trace elements	^{18}O and ^2H
	Heavy liquids	Major and trace elements	Not investigated
Squeezing		Major and trace elements	^{18}O and ^2H
Leaching		Major and trace elements	^2H and ^3H
Distillation	Under vacuum	Impossible	^{18}O and ^2H
	Microdistillation	Impossible	^2H only
	Azeotropic	Impossible	^{18}O and ^2H
Direct equilibration	With CO_2	Impossible	^{18}O only
	With water	Impossible	^{18}O and ^2H

3.1 Centrifugation

The use of a centrifuge to extract water from porous media started at the beginning of this century (Briggs and McLane, 1907). It was used mainly for establishing the "moisture equivalent" of a soil, that is the moisture content of a sample after the excess water has been reduced by centrifugation and brought to a state of capillary equilibrium with the applied force. However, early workers (Thomas, 1921; Joseph and Martin, 1923) observed a lack of homogeneity in the water content depending on the size of the samples, and Veihmeyer *et al.* (1924) observed that the moisture content of a specimen would increase with the distance from the axis of the rotation. Those discoveries led to the first attempts of a mathematical formulation of the driving force for water extrusion.

3.1.1 *Low-speed and high-speed centrifugation*

3.1.1.1 *Physical basis and design of the technique*

The centrifuge technique relies upon the pressure difference developed across the sample exceeding the capillary tension holding the water in the pores (Batlcy and Gilcs, 1979). The physics of fluid removal is fairly well understood, but the exact force distribution within the sample is difficult to determine.

Considering a column of water-saturated sediment under centrifugation, the applied tension T_a developed at a point within the column (r_2 [m] from the centre of rotation) will be given by the expression:

$$T_a = \frac{\omega^2}{2g}\left(r_1^2 - r_2^2\right)$$

where T_a is in metres of water; ω is the radial velocity in radians per second; g is the acceleration due to gravity [m sec^{-2}], and r_1 [m] is the distance from the base of the column to the centre of rotation.

The applied tension is consequently a function of the distance from the rotor and the speed of centrifugation. However, it will not depend on the density and the nature of the centrifuged material (Edmunds and Bath, 1976).

The removed interstitial water will depend on the pore-size distribution in the material. In fact, the capillary pressure in a pore will be:

$$T_c = \frac{4\sigma\cos\vartheta}{\rho d}$$

where T_c is the tension [Nm^{-2}]; σ is the surface tension; ϑ is the contact angle between the porous solid and the liquid; ρ is the specific gravity, and d [m] is the pore-channel diameter.

That formulation implies that it is not possible *a priori* to specify the size of the drained pores, since that parameter will vary within the sample. Edmunds and Bath (1976) (Figure 25) calculated some curves showing the relationship between rotation velocities and minimum drained capillary size, for the top and the middle of the centrifuged sample. In fact, for the balance between centrifugal and capillary forces, smaller pores should be emptied at the bottom rather than at the top of the column.

Figure 25. **Relationship between rotation velocities and minimum drained capillary size. Constructed assuming $\vartheta = 0°$, $\sigma = 0.07$ Nm^{-1}, $\rho = 1$, height of the sample = 6 cm, base of the column to the rotation centre = 11.1 cm. A = midpoint of the sample; B = top of the sample (after Edmunds and Bath, 1976)**

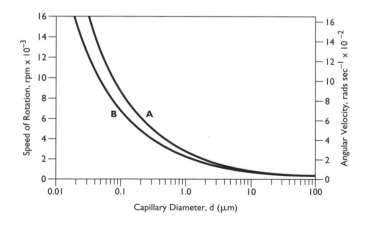

3.1.1.2 Conventional and basal cup arrangements: the distribution of water in the sample

For the same size grains, two possible cases must be taken into account: incompressible sediment grains and compressible sediment grains.

In the case of incompressible grains, if a conventional centrifuge tube is used, more water should remain where the centrifugal pressure is less (*i.e.*, at the top of the column). If grains are compressible, pores will be squeezed during the process, thus driving water to the top of the column. If that water is removed and centrifugation continues, a higher water content will result at the base of the column. That is why a basal cup arrangement (Figure 26) of centrifuge tubes (Schaffer *et al.*, 1937; Davies and Davies, 1963) may facilitate water drainage from the "bottom" of the sample, unless for very-fine-grained sediments where compaction may effectively block the water movement.

Figure 26. **Basal-cup arrangements (after Davies and Davies, 1963 and Edmunds and Bath, 1976)**

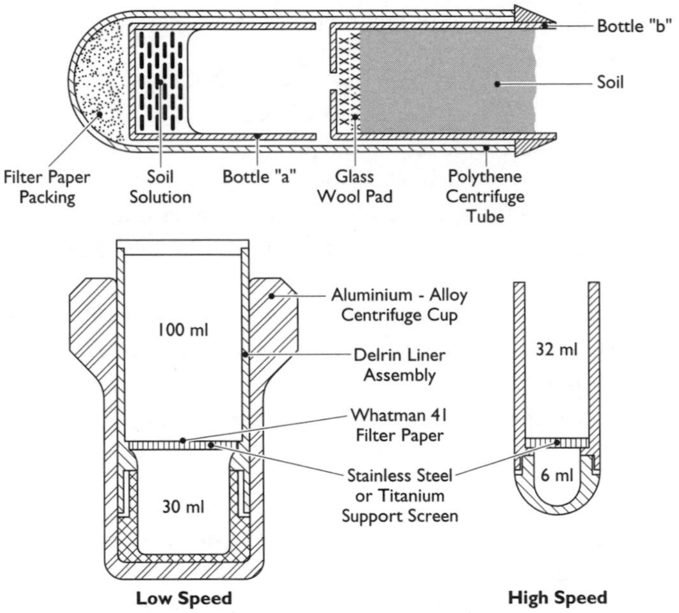

Batley and Giles (1979) compared the water distribution of different samples (sands, clays and mangroves) after centrifugation using those two centrifuge arrangements. They concluded that water collection at the top and/or the bottom of the sample depended on the specimen's grain size and compressibility, and may not be easily predicted.

3.1.1.3 Ultracentrifugation

With the advent of ultracentrifuges, the recoveries may be improved. On the other hand, however, ultracentrifuges commonly may treat only small amounts of samples and grain compaction is drastically increased. Batley and Giles (1979) found a small increase in recoveries for sandy sediments, while for clay sediments, the use of a high-speed centrifuge allowed to double the amount of recovered water.

Recently, that technique has been applied successfully for extracting water from unsaturated tuffs with fairly low-water content (7-30%) (Yang *et al.*, 1990). The extraction was performed both on intact samples (8,000 rpm, maximum weight of 780 g) and crushed samples (18,000 rpm, maximum weight of 115 g). The centrifuge is thermostatically controlled (25°C) and its chamber may be purged with dry nitrogen to prevent the sample from coming in contact with air when spinning. The percentage of recovery is approximately 75% for crushed samples and much lower (10%?) for intact cores (Figure 27). Chemical analysis, including pH, was performed on the collected water, and comparisons were made with the data obtained on similar samples by triaxial compression. A good correspondence, within 15%, was observed. The authors concluded that ultracentrifugation may provide samples from non-welded tuffs with initial moisture content greater than 11%. The use of intact cores is recommended, because core crushing exposes excess surfaces, eventually leading to a contamination of the sample if it comes in contact with air.

Figure 27. **Volume of water collected *versus* centrifugation time (after Yang *et al.*, 1990)**

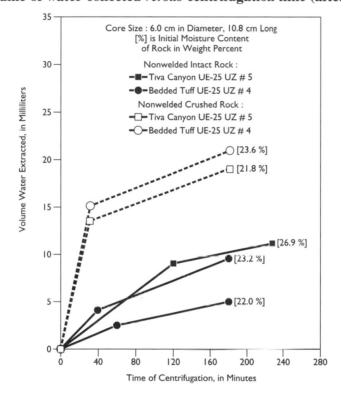

3.1.1.4   *Evidence of artefacts*

Since the beginning of the application of that technique, it appeared that it was necessary to prevent the evaporation of the sample during centrifugation by using closed tubes and to control the sample temperature for avoiding changes in porewater compositions (Schaffer *et al.*, 1937). Besides, even if it is not stated clearly, the impossibility to control redox conditions seems to limit the application of the technique to trace-element analysis and, more generally, to the extraction of representative porewater from anoxic samples.

Edmunds and Bath (1976) examined the composition change of the solutions extracted from chalk samples, as a function of the percentage of pore fluid removed (Figure 28). For sodium and potassium, they demonstrated a progressive depletion as a greater proportion of fluid is extracted, with an upturn or levelling-off over the last 10-20% (it should be noted that those trends are extrapolated from the general tendency for the last 10%). On the other hand, calcium showed a continuous depletion while magnesium and strontium showed no appreciable fractionation. As a possible explanation for that behaviour, they evoked a hydraulic drag effect on hydrated cations.[7] As their study was conducted on samples featuring a low percentage of clay minerals, they also noticed that the effect was unknown in lithologies not dominated by clays.

Figure 28. **Evolution of cumulative composition of extracted fluids after chalk samples (after Edmunds and Bath, 1976)**

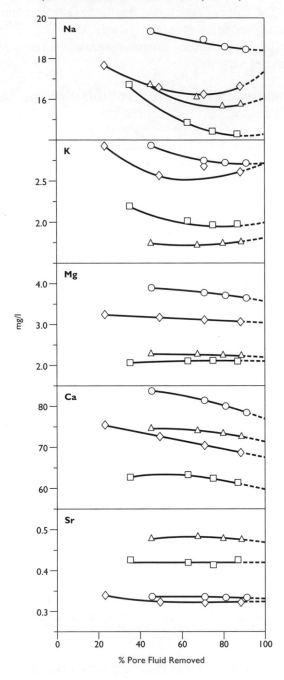

---

7.   A fluid moving against a rigid boundary exerts a dynamic force on it, caused by shear stresses and pressure-intensity variations along the surface. The component of that force in the direction of the relative velocity past the body is called the "hydraulic drag". According to the formulation given by the Stokes equation (Robinson and Stokes, 1970), the hydraulic drag would be much larger on divalent than on monovalent cations (Kharaka and Berry, 1973).

Recent studies (Zabowski and Sletten, 1991; Dahlgren *et al.*, 1997) showed the effect of carbon dioxide degassing on pore solution chemistry. Both studies were conducted on non-calcareous soils, thus minimising or controlling the artefacts induced by sampling and storage. They observed changes in dissolved organic carbon and major elements, leading to a general decrease in the ionic strength of the solution. The mechanism evoked is a cation adsorption on the clays for balancing the decrease on dissolved inorganic carbon (DIC) of the pore solution. On the other hand, minor changes were observed for the pH, probably controlled by other factors such as the buffering capacity of the clay minerals or the presence of organic acids in the soil solution.

3.1.1.5 *Centrifugation of frozen samples*

Within the framework of the ARCHIMEDE project, frozen samples were obtained by one of the involved teams (*Bureau de Recherches Géologiques et Minières – BRGM*). A test to verify if core freezing would increase water recovery during centrifugation was designed, based on the assumption that pore-size distribution should be modified in order to accommodate the ice-volume change. Samples of the frozen cores were carefully crushed trying to keep the pieces frozen, then placed in conventional closed centrifuge tubes and spun at 20°C for 2 h at 5,000 rpm.

No samples provided enough water to be collected. A thin mud film covering the centrifuge tubes would immediately solidify in contact with the atmosphere. The selected centrifugation parameters were probably not adapted. In addition, a water-content analysis of one of the samples used for the experiment prior to centrifugation showed that the parameter was sensibly lower than the normal one. Some problems of water-content modification (up to 30% loss) may have arisen from the collection and storage conditions (sublimation?) (see Chapter II § 2).

3.1.2 **Solvent displacement**

In 1976, Mubarak and Olsen suggested the use of a heavy immiscible liquid (CCl_4) in order to improve the recovery of the solution from dry-moist soils. The solvent (or "displacent") is put in contact with the solid sample in a conventional centrifuge tube. Since the solvent is immiscible with water and of greater density than water, the soil solution is displaced during centrifugation and rises to the surface where it may be collected for analysis (Figure 29). An attempt to describe mathematically the process is reported in Kinninburgh and Miles (1983).

Tests with different displacing liquids were performed, in order to verify any possible influence on porewater chemistry (Whelan and Barrow, 1980). Kittrick (1983) reported tests with four liquids (CCl_4, tetrachloroethylene, 1,1,1-trichlorethane and fluorocarbon FC70) and found them all suitable except for trichlorethane, significantly decreasing the pH of displaced solutions from calcite. Due to its toxicity, CCl_4 has currently been abandoned and Arklone (trichloro-trifluoro-ethane, density = 1.57) is the most commonly used (Kinninburgh and Miles, 1983; Moss and Edmunds, 1992).

Batley and Giles (1979) compared extraction yields obtained with low-speed and high-speed centrifugation, using or not using a displacent. They claimed a recovery increase of up to 5-10% for clay and mangrove sediments, but also noticed that the use of a heavy liquid as displacent may increase compaction in compressible sediments. Whelan and Barrow (1980) examined the retention curves displayed by different soils with variable clay contents (Figure 30). The authors showed a threshold water content below which no water may be displaced for a given set of centrifuging and displacement conditions. Above that threshold, there is a 1:1 relation between initial soil-water content and displaced

Figure 29. **Conventional centrifugation and heavy-liquid displacement**
(after Kinninburgh and Miles, 1983)

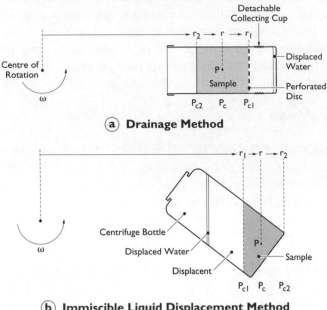

(a) **Drainage Method**

(b) **Immiscible Liquid Displacement Method**

water. The threshold depends on the texture of the soil (as reflected by the clay + silt percentage), the soil-water potential and the applied centrifugation.

Figure 30. **Water recovery from a range of soils with different initial water tensions**
(after Whelan and Barrow, 1980)

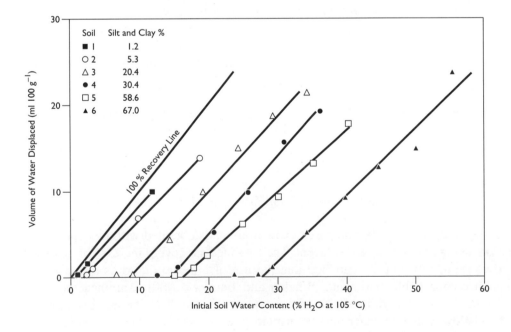

Whelan and Barrow (1980) also examined changes in solution composition as a function of centrifugation time. They detected only small changes for the considered species (NH_4^+, NO_3^-, P and Cl^-). Complete chemical analyses were not reported and tests were conducted on soils with more than 60% sand.

That technique proved better for controlling the redox conditions of the system for trace-element analysis and has been applied for that purpose in a number of studies (Batley and Giles, 1979; Kittrick, 1980; Kinninburgh and Miles, 1983; Moss and Edmunds, 1992; Edmunds *et al.*, 1992). On the other hand, a danger exists of extracting significant quantities of organic matter and associated metals from soils and waters. Batley and Giles found that the fluorocarbon FC78 extracted only 1-3% of DOC from groundwaters (and consequently associated metals), but that should be checked for every solvent prior to use.

3.1.3 *Stable-isotope analysis*

Only very few studies make use of the centrifugation technique for stable isotope analysis. Jusserand (1980) tested low-speed centrifugation, among other methods, for extracting water for oxygen-18 analysis. Reported results seem in fairly good agreement with data obtained by direct equilibration with carbon dioxide and vacuum distillation. The comparative study of Walker *et al.* (1994) (see Chapter II § 3.4.3) reported the use of high-speed centrifugation combined with Arklone displacement. Unfortunately, that technique was only able to provide a sample from a gypseous sand. The result showed no interaction with gypsum crystallisation water, but was affected by a rather high shift in isotopic composition with respect to dope water.

3.1.4 *Overall performance of the technique*

In the following table (Table 7), we have compiled some relevant porewater studies using centrifugation, in order to provide a synoptic summary of the performances of that technique.

3.2 Pressure filtering or squeezing

This section is extensively based on a recent review of the squeezing technique and its application to clayey sediments, carried out by the British Geological Survey (Reeder *et al.*, 1998) on behalf of GRS (Company for Reactor Safety, Germany). It has been integrated with some new data and comments mainly from CIEMAT (Energy, Environment and Technology Research Centre, Spain).

Pore-fluid extraction by squeezing or pressuring is the most widely used sampling technique for seabed sediments, especially those from water depth exceeding 1,000 m. It is also a method of extracting pore fluids from low-permeability clays and mudstones. Squeezing is analogous to the natural process of consolidation, caused by the deposition of material during geological times, but at a greatly accelerated rate.

3.2.1 *Theory*

The squeezing process involves the expulsion of pore fluids from the sediments being compressed. The sediments consist of solid particles (*i.e.*, the mineral phase) and spaces or voids that are filled with water in saturated rocks. When a squeezing stress is applied to a water-saturated material, its volume decreases due to three main mechanisms:

(i) compression of the solid phase;

Table 7. Case studies using the centrifugation technique

Reference	Method	Material	Initial water content	% of recovery	Chemical analysis / Notes
Sholkowitz, 1973	Not specified	Marine sediments	Not specified	Not specified	Major elements
Edmunds and Bath, 1976	Low-speed and/or high-speed centrifugation	Chalk	Not specified	~ 90%	Major elements
Mubarak and Olsen, 1976	Solvent displacement (CCl_4, 1,1,2-trichloroethane)	Clay soils (43% clay, 30% silt)	25%	50%	None
		Sandy loam (17% clay, 18% silt)	25%	52%	
Batley and Giles, 1979	Comparison of low / high speed centrifugation and solvent displacement (chloroform and FC78)	Clay pit	30%	38-40%	Heavy metals (with chloroform and FC78)
		Mangrove	48%	35-40%	
		Sand	23-29%	53-80%	
Whelan and Barrow, 1980	Solvent displacement (CCl_4, 1,1,1-trichloroethane, tetrachloroethylene)	Soils with different clay contents (0-47%)	12-65%	See text	NH_4^+, NO_3^-, P, Cl^- as a function of centrifugation time
Jusserand, 1980	Low-speed centrifugation	Marine sediments (85% clay, 10% $CaCO_3$)	40-75%	Not specified	Test for ^{18}O analysis (comparison with other methods)
		Lake sediments (30% clay, 40% silt, 20-25% $CaCO_3$ and 5-10% organic matter)	60-87%		
Kittrick, 1980	Solvent displacement (CCl_4)	Gibbsite and kaolinite	Not specified	Not specified	pH, Al (test for checking mineral solubilities)
Bath and Edmunds, 1981	High-speed centrifugation	Chalk	15-25%	40-50%	Major elements, Li, Br, F, ^{18}O, ^{2}H, but no alkalinity and pH (possibility of artefacts)

Table 7. **Case studies using the centrifugation technique (continued)**

Reference	Method	Material	Initial water content	% of recovery	Chemical analysis / Notes
Swalan and Murray, 1983	High-speed centrifugation	Marine sediments	Not specified	Not specified	Trace metals. T and O_2 controlled centrifugation
Kittrick, 1983	Solvent displacement (CCl_4, tetrachloroethylene, 1,1,1-trichloroethane, FC70)	Gibbsite and calcite	Not specified	Not specified	Test for mineral solubilities
Kinninburgh and Miles, 1983	Solvent displacement (Arklone)	Chalk	20% (wet)	80-85%	Major elements and heavy metals
		Soils	Variable	variable	
		Clay	20-40% (wet)	19-40%	
Zabowski and Sletten, 1991	Low / high-speed centrifugation	Spodsol soil	Not specified	Not specified	pH, DIC (test for CO_2 degassing effect)
Walker et al., 1994	High-speed centrifuge + Solvent displacement (Arklone)	Gypseous sand	10.2%	Not specified	^2H and ^{18}O (intercomparative with other techniques)
Moss and Edmunds, 1992	Solvent displacement (Arklone)	Sands, sandstones	8.3-12.5%	30-60%	Major elements, pH, Al
Edmunds et al., 1992	Solvent displacement (Arklone)	Sands, sandstones	Not specified	~ 50%	Major and trace elements
Leaney et al., 1993	High-speed centrifuge	Xeralf soils	20-40%	Not specified	^2H, cross check for azeotropic distillation
Yang et al., 1995	High-speed centrifuge	Non-welded tuffs	7-30%	Not specified	Some major elements
Dahlgren et al., 1997	High-speed centrifuge	Non-calcareous grey-wacke, sands, mudstones	Saturated	25%	pH, major elements (test for CO_2 degassing effect)

(ii) compression of the porewater between the solid phase; and

(iii) escape of water from the voids.

In most circumstances, the compression of the solid and liquid phases is negligible and most of the change in volume is caused by the escaping porewater. That may be illustrated by a hydromechanical analogy for load changing and squeezing, known as the "spring" analogy, as shown in Figure 31 (after Lambe and Whitman, 1979).

Figure 31. **A hydromechanical analogy for load changes during squeezing (after Lambe and Whitman, 1979)**

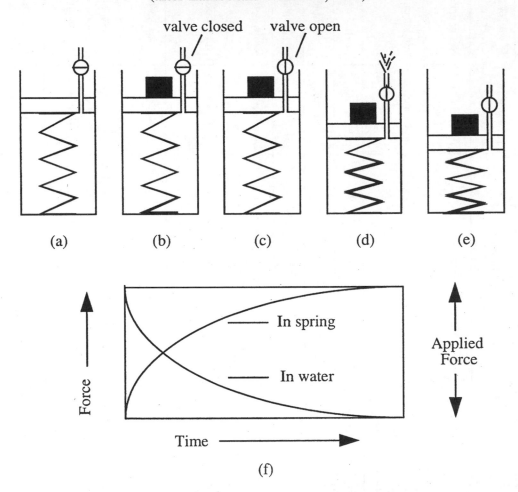

The strength of the solid phase during compression is represented by a spring and the rate at which the pore fluid flows depends upon the size of the valve aperture. In (a), the valve is closed and in equilibrium. When a load or stress is added (b), the piston load is apportioned by the water and the spring in relation to the stiffness of each. There is little movement in the piston because the water is relatively incompressible. Most of the load is carried by the water, and that increases the water pressure.

When the valve is opened (c), the excess pore pressure dissipates by water escaping through the valve (d). The piston drops and the volume of the chamber decreases until there is a new equilibrium when the applied load is carried by the spring and water pressure has returned to the original hydrostatic condition (e). The gradual load transfer from the water to the spring is shown in (f). The dissipation of the porewater is called "primary consolidation".

The change in volume is related to the stress applied and the difference between the stress and the pore pressure (Yong and Warkentin, 1975). The difference is the effective stress, σ', given by:

$$\sigma' = \sigma - u$$

where σ is the total stress and u is the pore pressure.

In non-fissured material, the expulsion rate of the pore fluid is related to the length of the sample and the pore size. A graph of the settlement rate after the addition of a load (Figure 32), and therefore of pore-fluid extraction, shows both primary and secondary consolidations.

Figure 32. **The settlement rate for increasing stress (Reeder *et al.*, 1998)**

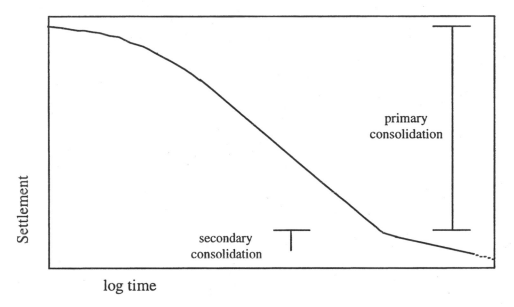

Most of the excess pore pressure dissipates during primary consolidation, while secondary consolidation involves the movement of particles as they adjust to the increase in effective stress and the dissipation of excess pore pressure from very small pores. The porewater extracted during squeezing is mainly due to primary consolidation.

The reduction in void volume (*i.e.*, pore size) with increasing stress reduces the permeability and, therefore, the rate of consolidation. The water extracted during each load increment or in total may be related to the change in void ratio, e, by the formula:

$$e = \frac{V_v}{V_s}$$

where V_v is the volume of voids and V_s is the volume of solids. Void ratio and porosity are related to porosity, ϕ, by:

$$e = \frac{\phi}{[1-\phi]}$$

Theoretically, at stresses above the yield stress (sometimes identified as the "pre-consolidation stress") the rate of change in void ratio decreases constantly with increasing log stress. Figure 33 shows void ratio *versus* log stress for an idealised unidimensional consolidation test.

Figure 33. Typical relationship between void ratio and log effective stress for an idealised unidimensional consolidation test of a clay material (Reeder *et al.*, 1998).

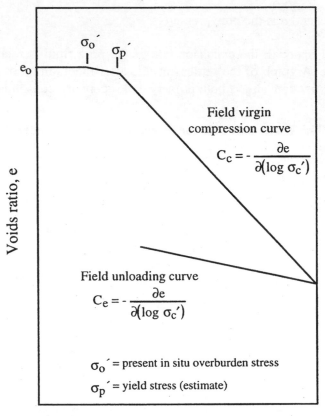

Effective consolidation stress, σ_c' (log scale)

The gradient of the virgin compression curve is the compression index, C_c, that may be related to the liquid limit, P_L (a commonly measured geotechnical parameter) by the following equation (Skempton, 1944):

$$C_c = 0.009 \, (P_L - 10)$$

That equation may be used to estimate the maximum volume of porewater that may be extracted, although the value of C_c is usually overestimated for overconsolidated clay and mudstones. Figure 33 also indicates that there may be little advantage in squeezing samples to very high stresses, as the increased volume of porewater extracted will be small.

3.2.2 Squeezing apparatus and practical applications

3.2.2.1 Overview

Pore fluids have been extracted from soils and rocks early on in this century. Ramann *et al.* (1916) and Lipman (1918) both used hydraulic squeezers. It was not until the 1940s when Kriukov

(1947) designed an easy-to-use, high-load compression device that pore fluids could be easily extracted from sediments. In the late 1950s, a filter press was adapted so that a direct pressure may be applied to the sediment (Lusczynski, 1961). Since then, two types of squeezing apparatus have developed: a low-pressure apparatus for sea or lake-floor sediments, capable of providing a maximum stress of about 1.5 MPa; and a high-pressure apparatus capable of squeezing fluids from a wide range of materials. The low-pressure types include gas pressure (Presley *et al.*, 1967), gas-mechanical action (Siever, 1962; Reeburgh, 1967) and hydraulic jacks (Kalil and Goldhaber, 1973). Most high-pressure systems use a hydraulic press and are similar to those designed by Kriukov (1947) and adapted by Manheim (1966). Those generated a maximum pressure of 5,000 kg cm^{-2}, equivalent to 480 MPa. A detailed review of pore-fluid extraction techniques, including pressure filtration and squeezing, was carried out by Kriukov and Manheim (1982).

3.2.2.2   Low-pressure squeezing apparatus

The low-pressure squeezing apparatus is specifically designed to extract water from high moisture-content materials such as marine and lacustrine sediments. Most modern designs are cheap and simple to use, thus allowing a number of tests to be performed at the same time. A detailed pore-fluid characterisation at a water/sediment interface may be carried out as a matter of routine with sample spacing of about 100 mm. Using non-metallic cells and filters minimises contamination, thus providing accurate measurements of trace elements. More recently, whole-core squeezers (Bender *et al.*, 1987; Jahnke, 1988) were designed for even closer sampling, and results with a resolution of 10 mm have been reported (Koike and Terauchi, 1996).

Early low-pressure systems were based on gas-displacement methods that have been used extensively to remove mud filtrate from oil-field drilling muds since their introduction by Richards (1941). Commercial mud presses use compressed gas from cylinders or cartridges to force filtrate through filter paper. Lusczynski (1961) used that method to study fresh/salt groundwater interfaces.

The extraction of porewater from seabed sediments was described by Siever (1962) who used relatively simple equipment. A filter press was modified to make it easy to load, clean and portable. The parts were all made of stainless steel except for the bronze piston and the neoprene O-rings and gaskets. The stainless-steel filter plate included many small holes and provided support for the filter paper. The load was applied by a turn screw attached to the upper plate or by a gas-driven piston, usually using nitrogen. The squeezer could accept a 76.2 mm-diameter core. The working pressure was between 700 and 1,400 kPa (100 to 200 psi) and between 20 and 30 ml of pore fluid were extracted from 100 g of 60-70% moisture-content clay in about 15 min. The composition of the fluid was not affected by those low-pressure conditions.

Hartmann (1965) described a gas-mechanical apparatus with a PVC membrane separating the confining gas from the sample. Nitrogen gas at a pressure of about 1,000 kPa (about 10 kg·cm^{-2}) passed through the PVC membrane, increasing pore-fluid recovery. In incompressible sediments, such as sands, the pore fluid was extracted mainly by the passage of gas but, in fine-grained clayey materials, the extraction was mainly due to compression of the membrane. The technique was used to extract about 95% of the water from a beach sand, yielding 50 ml in about 5 min. Finer-grained material from the Baltic Sea yielded up to 30 ml of water in up to 3 h; most of the pore fluid was extracted in the first 30 min. For those materials, about 40 to 60% of the water were extracted. Some of the pore fluids were cloudy and an extra filter was used after the water had come out of the press.

A non-metallic squeezer was developed for trace-metal studies at the Department of Oceanography of The John Hopkins University, in Baltimore, Maryland, USA (Reeburgh, 1967). It has no moving parts and, like the Hartmann type, the gas pressure acts on a diaphragm compressing the sediment and forcing out the porewater through filters into a sample bottle. The sample cell is about 50 mm in height and 50 mm in diameter, and all the components are made of Delrin, nylon or dental-dam rubber. The cell is held together with a slightly modified C-clamp. The resulting apparatus is easy to clean up, but the gas pressure is limited to 1,400 kPa. A single gas cylinder has been used to extract porewater from up to 20 cells simultaneously via a manifold (Kriukov and Manheim, 1982). Nitrogen, carbon dioxide and helium have been used as the pressurising gas. The pore fluid extraction was carried out using gradual increases of pressure over about 30 min and about 25 ml of pore fluid per 100 g of sediment were recovered. Since the system does not contain metal parts, it is ideal when investigating trace metals and labile constituents.

A system slightly modified from the Reeburgh-type squeezers was described by de Lange (1992). In that case, the supporting screen was made of Teflon with 0.2-mm cellulose-acetate membrane filter. The membrane was glued to the O-ring providing a quick and secure sealing system. The supporting screen reduces dead volume. The squeezers were fixed with slide-rod clamps, and three-way stopcocks allowed the squeezers to be used independently. The tests were carried out in a nitrogen anaerobic box. The working pressure was applied gradually up to 1,500 kPa. Squeezing times varied between 15 min and 4 h depending on the type of sample. About 150 cm^3 of sediment were squeezed and were sufficient to produce between 15 and 60 ml of pore fluid. Up to 300 cm^3 of sediment may be handled, if required.

Patterson et al. (1978) described a non-metallic mechanical squeezer based on the Kriukov-Manheim design to be used on compressible, fine-grained or organic-rich sediments. The maximum stress applied was 5 MPa.

3.2.2.3   Whole-core squeezers

Whole-core squeezers were designed to reduce handling time and to minimise core disturbance and the introduction of oxygen to anoxic core samples. The different apparatus devised by Bender et al. (1987) and Jahnke (1988) allow for a very detailed sampling. The Bender system was successfully applied by Koike and Terauchi (1996) to assess the fine-scale vertical profiles of nitrous oxide in near-shore sediments.

The basic design of the instrument detailed in Bender et al. (1987) comprised a lucite piston with a narrow hole through the centre, a filter or fine mesh screen (10-μm polyester) taped to the bottom of the piston and an O-ring in grooves on the side of the piston to provide a seal. A stationary piston with an O-ring sealed the bottom of the core liner. Stress was applied via a rod attached to the upper piston with a vice. The upper piston was placed above the sediment column in the bottom water sampler. As the piston was lowered, water was forced out into a sampling tube. At the water/sediment boundary, the water extraction continued, but pore fluid was extracted this time. The water was collected in approximately 3-ml aliquots that equated to water from every millimetre, thus providing good resolution. Pore fluid from the top 20 mm typically took between 30 to 60 min to collect, but below that level, extraction was very slow due to the compaction of the sediment.

Several problems may be expected with that technique, such as contamination of the sediment by the bottom waters, internal mixing of the porewaters during extrusion, and solid/solution reactions as the pore fluid is expressed through the overlying sediment. Bender et al. (1987) detailed the techniques used to test each of those possible artefacts. Tracer experiments indicated that any mixing associated with squeezing did not smear profiles by more than a few millimetres. However, the squeezed porewaters were very susceptible to rapid alteration by solid/solution reactions, and it was concluded that the technique was not suitable for trace metals and other particle-reactive chemicals.

The whole-core squeezer devised by Jahnke (1988) was similar in concept to that of Bender *et al.* (1987), except that the pore fluids were expelled out of a number of sampling ports when a little pressure was added. That technique was devised to be used at sea and allowed the sampling of pore fluids from as little as 10-mm spacing without sectioning the core.

Holes at the required sampling distance were drilled into an acrylic core barrel, tapped and sealed with nylon screws. The device was inserted into the sample core box and a subsample was taken. The ends were capped with O-ringed pistons so that there was sufficient water above the sediment to prevent any air from being trapped. The cell was placed into the pressuring system comprising two angle braces attached to a unistrut and two threaded rods with nuts screwed directly into the pistons. Pressure was applied by screwing the threaded rods into the nuts, thereby moving the pistons into the core barrel. The lower nut was turned until the sediment surface was at the desired sampling level relative to the sampling ports. Once the core was in position, the nylon screws were removed and replaced with luer fittings equipped with a 0.45-μm filter and plastic disposable syringe. The bottom or top nut was then periodically turned to pressurise the core, forcing out pore fluid into the syringe. Alternatively, the core may be pressurised with gas. A pressure of 300-400 kPa was generally used.

The results of that technique were compared with those of an adjacent core that was sectioned and centrifuged in a glove box with a nitrogen atmosphere. The results were similar apart from the iron profile and the lower surface values of phosphate, silicate and iron for the squeezed test. The latter is probably due to the movement of the sediment relative to the sampling points during squeezing. That discrepancy may be reduced by squeezing from the top and the bottom, thus reducing the 1-2-mm smearing of the profile. However, the apparatus was easy and inexpensive to build and the technique less invasive than sectioning. With proper care, there was no need for glove boxes, and the assembly could be fitted into a large refrigerator. Experience showed that about 10 ml of pore fluid may be extracted from 10 depth intervals in less than 1.5 h, and that several may be performed simultaneously.

3.2.2.4   Heavy-duty squeezing apparatus

Heavy-duty squeezing apparatus is capable of recovering pore fluid from stiff, firm or even hard, low-porosity materials. Early on in this century, "hydrostatic presses" were used to extract pore fluid from soils (Ramann *et al.*, 1916; Lipman, 1918). In the 1940s, a practical stainless-steel squeezing device with a self-sealing free gasket was designed by Kriukov (1947) and used for agronomic and geological purposes. More recent applications to soil-solution studies were reported in Heinrichs *et al.*, 1996 and Böttcher *et al.*, 1997.

Manheim (1966) described a hydraulic squeezer based on that of Kriukov with a few alterations. The apparatus was constructed mainly from commercially available parts and used a standard press. It was designed to extract water from small quantities of mostly Recent to Palaeocene sediments. Extraction took as little as 3 min. Similar equipment has been used by many others, including: Jones *et al.* (1969), to extract water from saline lake and playa deposits at Abert Lake, Oregon; and Morgenstern and Balasubramonian (1980) to investigate the swelling characteristics and the chemistry of pore fluids extracted from Bearpaw shale in Saskatchewan and Morden shale in Manitoba (Canada).

That squeezing apparatus had a cylinder and ram construction, using a standard 10-t load frame and press to apply the load. The original squeezer used a metal cylinder of 28.6-mm inside diameter, 50.8-mm outside diameter and 76.2-mm height. The cylinder fitted onto a machined base. The filter units comprised stainless-steel screens and a perforated steel plate or a sintered disc fitted into a filter holder. The seals at the base and ram end were provided by a rubber or neoprene washer with a Teflon disc at the top. All the metal parts were manufactured out of ISI No. 303 stainless steel. The 10-t press could exert a maximum pressure of about 150 MPa on the sediment. A larger squeezer with an internal bore of about 57 mm was also constructed with a thicker filter plate for greater strength. The maximum stress on the sample using the 10-t press was about 35 MPa.

A larger version of that design was manufactured for the British Geological Survey (Brightman *et al.*, 1985) to obtain porewater samples from the Mesozoic mudstones at the Harwell Research Site. Those compacted mudstones had typically 14 to 25% moisture content by dry weight. The cell was 75-mm inside diameter and could accommodate samples up to about 100 mm long. The maximum working load of the cell was supposed to be 200 MPa. Since calculations by the Atomic Energy Research Establishment showed that stainless steel did not have the required characteristics, a nickel/chrome/molybdenum alloy (BS970/PE2 826 M40) was used.

The apparatus based on Manheim (1966) required a number of modifications. The neoprene seals extruded under high loads and were replaced with perbunam rubber seals. Unfortunately, those still extruded through the drain hole at stresses greater than 110 MPa. During early tests, particles were extruded from the sample through the drilled plate and mesh filter that was replaced with a sintered bronze filter disc. Those filters were used only once as they deformed under load. The filter support was changed to a flat upper plate easing the removal of the deformed filter. The base plate was changed to stainless-steel one after the original base plate showed signs of corrosion. A large concrete 2,000-kN press provided the load. Test specimens varied between 430 and 740 g, and most of the samples had moisture contents varying between 10 and 29%. Porewater flowed through a 0.45-μm Acrodisc© filter and collected in plastic syringes. The quantity of pore fluid extracted depended on the type and size of the initial sample, its moisture content and the amount of water required. Most tests produced between 30 to 50 g of pore fluid within 5 to 7 d.

The Kriukov-Manheim design was modified by Nagender Nath *et al.* (1988) by the addition of a Teflon inner-cell liner to ensure that the sample and pore fluid did not come in contact with metal. The maximum volume of the sample chamber was 80 cm^3. A system of perforated Teflon filter discs with offset holes was used at the bottom of the sample to overcome the movement of sediment through the filter paper. The base of the squeezer had a vertical slot in which the sample collectors (plastic vials) were housed so the pore fluid did not come in contact with the atmosphere.

Fontanive *et al.* (1985) used a Kriukov/Manheim type cell and an oedometer press for loads up to 3.3 MPa to extract pore fluids from Italian Plio-Pleistocene formations. The cell containing the sample was then transferred to a hydraulic press with a maximum stress of 104 MPa. The samples were 50 mm in diameter and had a maximum height of 45 mm. Initial moisture contents of the samples varied between 11.7 and 34.9%. Samples with initial moisture contents of 29.6% and 19.1% produced 23.16 and 12.00 ml, respectively.

A new cell was developed at the British Geological Survey in the late 1980s (Entwisle and Reeder, 1993) to extract pore fluid from glacial tills as well as Mesozoic and Tertiary mudstones. A wide range of chemical tests was required, including minor elements and isotopes. Relatively large quantities of pore fluids were thus needed: up to 80 g per test from materials with moisture contents of about 20%, and reasonable quantities of at least 20 ml from materials with moisture contents around 10%. Experience from the earlier Kriukov-Manheim type apparatus (Brightman *et al.*, 1985) indicated that large samples of up to 800 g and long test times were required.

The new squeezing apparatus used commercially available hydraulic pumps with a maximum output stress of 70 MPa and a capability to maintain constant stress over a number of weeks. A 50-t long-stroke hydraulic ram transferred the stress to the sample and provided a maximum working pressure of 70 MPa. The long stroke allowed the test sample to be extruded from the cell afterwards.

The main body of the cell and other metal parts in contact with the test sample or pore fluid were made of Type-316 stainless steel that was selected for its resistance to corrosion and high tensile strength. The sample chamber was 75 mm in diameter and 100 mm in height. A spiral groove machined into the outside of the cell was covered with a plastic jacket to take a temperature controlling fluid, thus enabling squeezing tests to be conducted under controlled temperatures between 0°C and 50°C. A schematic diagram of the cell is presented in Figure 34.

The bottom platen of the apparatus screwed to the top of the hydraulic ram was sealed with 12V Pack Seal™ containing a nylon packer in order to stop any fluid leakage from the bottom of the sample and to scrape clean the inner wall of the cell as it moved up. The top platen screwed to the top of the cell and was sealed with an O-ring. A separate pore-fluid pipe screwed to the top platen was sealed with an O-ring. That enabled easy cleaning access and reduced the dead volume. A paper filter was placed onto the sample and supported with a stainless-steel filter. The apparatus may be used with an anaerobic chamber, thus enabling samples to be prepared and squeezed in a very low oxygen atmosphere (minimum 2 ppm of oxygen).

The apparatus has been used to extract pore fluid from materials with moisture content slightly below 7%. Only a few grams of pore fluid may be extracted from mudstones with less than 9% moisture content. Unsuccessful tests tend to concern highly cemented, hard material. Porewater obtained using that apparatus has been analysed for up to 51 chemical species and the isotopes of seven elements.

Yang *et al.* (1995) described the pore-fluid extraction from welded and non-welded tuffs using a high-pressure unidimensional compression apparatus for samples with initial moisture content greater than 6.5% and using a triaxial system for samples with greater than 8% moisture content.

The unidimensional compression apparatus was based on a compression machine designed for research on concrete (Barneyback and Diamond, 1981). It was simple to operate and provided a wide range of stresses. The main components were made of 4340-alloy steel. The inner-sample sleeve and the drainage plates were made of Monel K500 nickel alloy. Core samples were wrapped in a Teflon sheet and confined in the sample tube. The sample chamber was sealed with O-rings and a Teflon washer. Fluids were collected in syringes at both ends of the rig. That closed system prevented any contact between the expelled water and the atmosphere. The maximum sample size was 61 mm in diameter by 110 mm in length. Samples shorter than 55 mm were accommodated by inserting steel spacers beneath the base plate. The apparatus had a maximum stress of 552 MPa and required a compression machine with a capacity of 1.6 MN. Stresses were added in stages. It was found that the porosity decreased by 36 to 79% for non-welded tuffs and by 25 to 49% for densely-welded tuffs.

The triaxial compression machine was modified from a standard, commercially available triaxial cell. The cell and end caps were made of 4140-alloy steel and the lateral stress was applied through a urethane membrane. The maximum sample size was 61 mm in diameter by 113 mm in length.

Figure 34. **Schematic diagram of the squeezing cell described by Entwisle and Reeder (1993)**

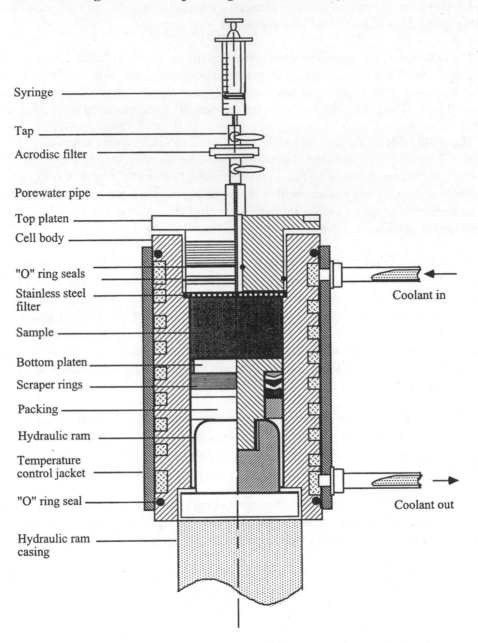

Syringe

Tap

Acrodisc filter

Porewater pipe

Top platen

Cell body

"O" ring seals

Stainless steel filter

Sample

Bottom platen

Scraper rings

Packing

Hydraulic ram

Temperature control jacket

"O" ring seal

Hydraulic ram casing

Coolant in

Coolant out

The fluids were collected at both ends of the rig in plastic syringes. A hydrostatic stress of 68 MPa was applied and then a maximum axial stress of 193 MPa was applied in four stages.

Dry nitrogen was injected into the sample after fluid had been collected from the last loading stage for both techniques described by Yang *et al.* (1995). Pressures varying between 0.3 and 9.7 MPa were used for the triaxial test and between 1.4 to 10.3 MPa for the unidimensional test. The injection continued until no more water was forced out. The unidimensional method produced more water and from lower moisture content samples than the triaxial method. The porewater compositions from both triaxial compression and high-speed centrifugation were comparable.

A system similar to the unidimensional squeezing apparatus described by Yang *et al.* (1995) was used by Cuevas *et al.* (1997) on bentonite from Serrata (Almería, Spain). The compaction of the sample was carried out by a hydraulic press that provided a maximum stress of 3,000 kN. The compaction chamber was made of type AISI 329 stainless steel (due to high tensile strength and resistance to corrosion). The maximum sample size was 70 mm in diameter and 118 mm in length. A maximum working pressure of 60 MPa was applied to the sample and that stress was maintained constant over a number of weeks. The drainage system allowed for the extraction of porewater at each end of the sample through a 0.45 mm sintered AISI 316L stainless-steel filter disc, and the porewater was collected in a polypropylene syringe.

That squeezing apparatus was designed to study the behaviour of a Spanish bentonite subjected to simultaneous heating and hydration in a thermohydraulic cell. A sample of bentonite compacted at 1.62 g/cm^3 and 11.2% of water content was simultaneously heated at 100°C at the top and hydrated at the bottom of the sample for three months. At the end of the test, the compacted sample was taken out and sliced in five horizontal sections (section 1 being the closest to the heater and section 5, the closest to the water source). The final temperature reached ranged from 100 to 60°C and the final water content ranged from 14 to 24%. Each of the sliced samples (700-900 g) was introduced in the squeezing chamber and compacted at 64 MPa to study the changes in porewater composition due to temperature gradient and water saturation, the transport mechanism of salts and possible precipitation/dissolution processes. Porewater was only obtained from three sections (those with a water content over 20%) at a pressure of 64 MPa.

It was proven that all squeezed bentonite samples always reached a final dry density of about 1.72 g/cm^3 at a pressure of 64 MPa. The water collected corresponds to the amount of excess water compared to that of saturated clay at a density of 1.72 g/cm^3 for a 64 MPa pressure.

A new squeezing chamber (type AISI 329 stainless steel) was designed by CIEMAT to study the changes in solution chemistry with applied pressure. That cell endured a working pressure varying from 25 to 200 MPa. The water coming out of the compaction chamber was first channelled in a Teflon tube and then passed through different microelectrodes, where pH, Eh, electrical conductivity and temperature were measured and the values stored in a multiparametric data-acquisition system. Finally, the water was collected in a syringe (Fernández *et al.*, 1996; in preparation).

3.2.2.5 Whole-core sampler and squeezer

A robust whole-core sampler was designed for field work at the British Geological Survey (Kinniburgh *et al.*, 1996). It successfully extracted pore fluids from partially saturated chalk, for biological-oxygen-demand (BOD) analysis. The sampler was an especially prepared, steel U100 core barrel with small water-sampling holes drilled into the side and sealed with a grub screw. A drill rig drove the U100 into the chalk. Upon return to the surface, the U100 containing sample was placed into a horizontal-load frame with a 30-t, long-stroke hydraulic jack. Pistons with O-rings were placed at each end of the sample. The stress was applied with a 25-t hydraulic pump providing a maximum stress of about 30 MPa. The water was collected via a screening system into plastic syringes.

3.2.3 Squeezing technique

3.2.3.1 Porewater extraction

A typical working procedure for squeezing marine samples with pressure-filtration systems is detailed in de Lange (1992). After core collection, the PVC liner is cut into 1-m sections, tightly sealed and stored horizontally at 3 to 4°C. The pore fluid is extracted within 24 h of core collection. The core-section is split lengthways into two parts: one is transferred into an anaerobic glove bag, and small subsamples are taken from the other for various tests, including moisture content. The samples to be squeezed are put into beakers and placed in a glove box. The sediments are transferred into the pressure-filtration cells. Squeezing starts at low pressure, that is gradually increased up to 1.5 MPa in order to reduce the possibility that a dry sediment cake forms above the filter. The test rate and the fluid volume produced depends upon the type of sediment and its initial moisture content. After the required amount of water has been squeezed out of the sample, the sediment is subsampled to determine the remaining moisture content.

Sample preparation is important to prevent oxidation and other disturbances. For tests where oxygen-sensitive species are important, core handling should be carried out in a controlled anoxic environment. At the British Geological Survey in Keyworth, samples are prepared and tested in an atmosphere of about 100 ppm of oxygen or lower. The oxygen content of the chamber usually rises when the core samples are opened.

Where possible, at least 10 mm should be removed from the outer annulus of the core material prior to testing to reduce the possibility of contamination with drilling fluids. Disturbed surface samples received in bags may be tested directly. Stiffer core materials, such as overconsolidated clays and mudstones, may be prepared as cylinders in a soil lathe using a knife, soil saw or scalpel. Harder, more indurated materials are broken up with a sturdy knife and hammer. The outer part of the core is removed and the inner material is placed into the cell. Clean disposable gloves should be worn during sample preparation.

The initial weight of the squeeze sample should be measured in order to determine the extraction efficiency. The moisture content of the sample should also be determined using off-cuts collected during the sample preparation. Clays and mudstones usually take a few days to squeeze and the initial moisture content may serve as a useful guide to the stress regime, or, in the case of low-moisture content samples, a guide to whether the test is likely to produce porewater.

Operation of the Kriukov-Manheim-type squeezing apparatus is as follows. The filter holder with its rubber or neoprene gasket is placed in the recess of the filter base. The porous disc or screen and filter papers are placed on top. The sample, of known weight, is rapidly transferred into the cylinder. The upper seal, Teflon and rubber disc are placed on top of the sample and the piston is placed into the cylinder and depressed as far as it will go. The unit is transferred to the press. Larger cells are assembled in the press. Pressure is applied gradually until the first drops of water are expelled, then a syringe of known weight is placed into the pore fluid port. When the required amount of liquid has been collected, the syringe is removed, weighed and capped. The syringe is then placed into a refrigerator prior to preservation of the porewater and subsequent chemical analysis. If a number of pore-fluid samples are required, a syringe tap is pushed into the port in order to be able to stop the flow of liquid while syringes are being changed.

Pressure is added gradually so that fluids are not extracted too rapidly in order to prevent the production of a cloudy liquid containing colloids or even clay particles. It may take from a few days to a few weeks to produce pore fluid from moderate to low moisture-content clays and mudrocks. When several subsamples of pore fluid are required from the same squeezed sample, the extraction time of the later samples increases as the hydraulic conductivity of the sample decreases.

The squeezing temperature may be controlled by conducting the test in a temperature-controlled room or with a temperature-controlled fluid circulating between the cell and an insulating jacket.

After the test is completed, the squeezed sample is removed from the cell, weighed and usually oven-dried in order to ascertain the original moisture content of the test specimen. Based on that information, it is possible to calculate the water content of the sample at the start and the end of the test. The moisture content may also be calculated before and after each fluid sample is taken.

The record of the pore-fluid extraction should include information on the apparatus, the test sample, the test method, the mass or volume of water extracted and the stress history of each extraction.

3.2.3.2. *Determination of moisture content and extraction efficiency*

Duplicate samples of the geological material should be taken at the same time as the sample is prepared for the determination of moisture content. At least 50 g of original sample should be tested by determining its exact weight before and after heating at approximately 105 to 110°C for a minimum of 24 h.

The moisture-content percentage with respect to dry weight, M_w, is determined as:

$$M_w = \frac{\left[W_w - W_d\right]}{W_d} 100\%$$

where W_w is the weight of the wet sample, and W_d is the weight of the dry sample. Reporting moisture content with respect to dry weight is standard in squeezing tests.

The percentage of the available porewater extracted during squeezing, E, (the "extraction efficiency") may be determined by:

$$E = \frac{W_p}{W_{si} - W_{sd}} 100\%$$

where W_p is the weight of the collected porewater, W_{si} is the weight of the sample initially tested, and W_{sd} is the post-squeezing weight of the sample after oven-drying at 105 to 110°C for a minimum of 24 h.

3.2.4 *Squeezing artefacts*

The effects of pressure filtration were considered early in the development of the technique. Northrup (1918) thought that high stresses may affect the physico-chemical equilibrium in the soil. In some experiments, the ionic content of the pore fluid increases (Kharaka and Berry 1973). However, in tests on sodium-chloride solution mixed with Askangel clay and Oglalinski bentonite, Kriukov *et al.* (1962) observed marked decreases in solute concentration with increasing pore-fluid extraction. The decreases started at about 70% moisture content for the Askangel clay and at about 47% moisture content for the Oglalinski bentonite. Squeezing ion-exchange columns containing the sodium-chloride solution gave similar results. The pore-fluid chemistry of a saline mix with Glukhovetsk kaolinite showed little change when extraction increased. It was thought that the high-salinity water was removed from the clays at high moisture content and the lower concentration of the later samples was due to the lower-salinity solution remaining in the compacted clays.

Extensive studies of pore-fluid chemistry on marine sediments indicates good reproducibility (Manheim, 1974). Pore-fluid chemistry of marine sediments is very sensitive to sampling, storage and test method. Manheim (1974) considered that the variability of results reported for interstitial solutions from the Gulf of Mexico (Parashiva Murthy and Ferrell, 1972; 1973) arose from problems of technique. Absolute composition of interstitial solutions from sediments near the sediment/water interface may be compared to that of seawater. Pore fluid from sediments just below the water/sediment interface ought to have similar composition to the local bottom water. For chloride, agreement may be within 1%. The understanding of the diagenesis of minerals in marine sediments is aided by information about changes in pore-fluid chemistry at depths below the seabed. In marine core, the pore-fluid chemistry usually changes gradually and smoothly with depth due to diffusion (Manheim, 1974; Bender *et al.*, 1987; Jahnke, 1988). The porewater composition changes with depth below seabed; sodium and magnesium tend to decrease, and calcium increases. The response of different chemical species varies with the type of sediment, the organic content and the deposition rate and diagenesis. Boron was found very temperature-sensitive, but that may be compensated for.

Squeezing tests in overconsolidated clays, mudstones and tills have also shown smooth concentration profiles across the formation for the more conservative species, especially chloride (*e.g.*, Bath *et al.*, 1988; Metcalfe *et al.*, 1990; Reeder *et al.*, 1997). Increases in calcium, magnesium, sulphate and a number of other ions have indicated the presence of gypsum or higher concentrations of oxidising pyrite. Recent work has shown that the diffusion patterns down a hillside of Pliocene mudstones in Italy was a function of solute movement driven by surface evaporation. Comparison of pore-fluid chemistry of chalk from squeezing and centrifugation showed many similarities (Bath *et al.*, 1988). Tests carried out on adjacent samples of London clay, with an initial moisture content of about 19%, showed good reproducibility when calculated back to the moisture content of each pore-fluid sample.

Manheim (1974) suggested that the threshold of pressure influence above which there was a reduced ionic strength should be established. Chilingarian *et al.* (1973) observed that the concentrations of porewaters squeezed from montmorillonite clay were slightly higher than the porewater present during the early compaction stages, but then started to drop when the overburden pressure increased. Morgenstern and Balasubramonian (1980) similarly observed a reduction in total cation content when stress rose beyond 20.7 MPa. That was considered to be due to dilution by water from the double layer, and by low-porosity clayey sediments exhibiting a membrane effect.

For lower moisture-content clays and mudstones, sequential pore-fluid samples show changes in pore-fluid chemistry (Entwisle and Reeder, 1993). For many samples, there is a decrease in monovalent species with increasing pore-fluid extraction and an increase in divalent and then trivalent species. The stress level used to extract the pore fluid may be an indicator of changing chemistry, but the change is more likely to be due to the reduction in moisture content, pore size and void ratio as the particles move closer together. Recent work at the British Geological Survey using both squeezing and leaching techniques suggests chloride exclusion from the micropores of very low-porosity mudstone.

3.3   Leaching

3.3.1   Principle of the technique

Multiple washing of soil specimens and sediments was at first designed to obtain information on the ions that may be released by the solid phase (Gedroiz, 1906, Cameron, 1911; Parker, 1921). In 1969, Mangelsdorf *et al.* suggested the use of such a technique to evaluate the artefacts induced by temperature changes over porewater chemistry in a seawater/montmorillonite mixture. Schmidt (1973) and Lisitsyn *et al.* (1984) also suggested the use of that technique to obtain information on groundwater composition from sidewall and borehole cores.

Parshiva Murthy and Ferrell (1972; 1973) started from the observation that porewater squeezed out at 100 psi was enriched in ionic constituents relative to the solution obtained by a multiple-washing technique. For their experience, they used freshwater sediments and freshly boiled deionised water. The sediments were mixed with a varying amount of water (1:2; 1:5; 1:10) and the two phases separated by centrifugation. They observed an increase in the calculated concentrations of cations in the water extract with a decrease in the sediment/water ratio. Besides, since monovalent ions would increase more than divalent ions, that result would be in general agreement with the prediction of the Donnan principle. They also stated that the "in-place" values of ions in the interstitial solution may be derived by the extrapolation of the dilution data to the observed moisture content of the sediment sample, the Donnan equilibrium being the major control on the distribution of ions in the clay/water system.

A more detailed theoretic approach to leaching is reported in Devine et al. (1973). In their opinion, the clay/water system is composed of an inner solution that is in contact with the clay particle and an outer solution. The exchange between both solutions is due to Donnan-equilibrium interactions. In order to obtain the composition of the outer solution, several aliquots of the solid sample should be diluted with deionised water, and the solution separated by centrifugation. During washing, both the outer and the inner-solution compositions would be affected, while the activity of the metal ion on the clay particle would remain constant (Figure 35). In order to obtain reliable results, sediment samples without readily-soluble salts should be selected. In-situ concentrations would then be obtained by extrapolating the ion concentrations of the leachates to the in-situ water content of the specimen (Figure 36). The authors suggested, as also shown by Mangelsdorf et al. (1969), that the relationship log (water/sediment) to log (activity of the ion) should be linear. According to Devine et al., the method is suitable for obtaining the concentration of the ions in the outer solution. The composition of the inner solution should be obtained by extracting the total pore solution from an unprocessed aliquot of the sediment (no method is specified) and calculating it by difference with the outer solution.

Schmidt (1973) used leaching to derive the porewater composition of shales, using sidewall cores. In his procedure, samples were oven-dried, crushed, placed in a stainless-steel centrifuge tube with 50% ethanol/distilled-water solution. The sample was dispersed ultrasonically for 1 h and then ultracentrifuged. The extraction procedure was repeated three more times, combining the extracts. In addition, the author carefully measured the exchange capacity of shales as a function of depth for sodium, potassium and lithium (CEC ranging from 140 to 20 meq/100 g). Despite those measurements, as there was "a close agreement of the milliequivalents per litre of the cations and an(t)ions", he concluded that cations were actually soluble cations and not cations on the exchange sites of any excess clay in the solution. Hence, the extraction procedure is valid for estimating soluble salts in shales. From the discussion, it appears that the author had not a clear idea of the significance of the cation-exchange process. In addition, the electroneutrality of the extracted solutions is in no way a means of detecting released cations from the solid and only indicates the error in the analysis.

Leaching has not been investigated in detail for isotopic analysis, as it requires labelled compounds. Zimmermann et al. (1967) reported on the use of that technique combined with centrifugation to obtain deuterium profiles in the unsaturated zone. The use of that isotope involves a conversion from weight per cent to volume per cent of the original moisture content. That step is not trivial, because the conversion needs either the bulk density of the soil profile (that may be easily disturbed during sampling) or the volume of the solid sample. The possible error is estimated to be lower than 10%. Hence, tritium is used instead, because it is much more inexpensive and may be measured by liquid-scintillation counting, thus avoiding the conversion step.

Figure 35. **Processes occurring at the vicinity of a clay surface during dilution of the interstitial solution (after Devine *et al.*, 1973)**

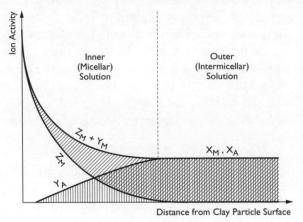

(a) Unprocessed Sediment

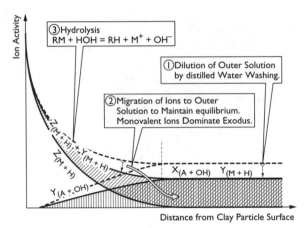

(b) During Washing

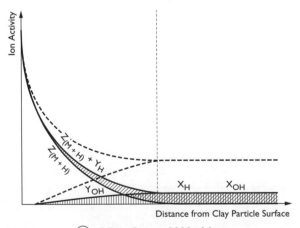

(c) After Several Washings

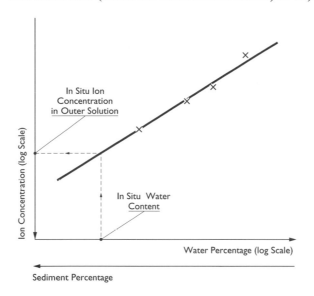

Figure 36. **Graphic method for determining the outer solution concentration. The original Figure has been modified because the graphical construction was incorrect (modified after Devine *et al.*, 1973)**

3.3.2   *The problem of the solid/water interaction*

Since its publication, the work of Parshiva Murthy and Ferrel (1972) was strongly criticised, both for its approach and for the total lack of supporting data. That led the authors to publish in 1973 a compilation of the data obtained by leaching with deionised water (providing interstitial ions), by leaching with a $1N$ ammonium-acetate solution (providing exchangeable cations), and by acid attack on NH_4-saturated samples (providing non-exchangeable cations). They suggested that the multiple-dilution treatment was the best method to obtain the distribution of the ions between the interstitial and exchangeable positions in sediments, and should be adopted as a uniform standard to report results of sediment-water analysis. As reactions other than dilution and dissolution of sparingly-soluble salts are involved at high water/sediment ratios, thus resulting in a rearrangement of exchangeable and non-exchangeable ions, the abundance of interstitial ions should be determined by a one-step dilution method. Despite that clarification attempt, the concept of their approach is, in our opinion, confused and based on wrong premises.

Walters (1967) found that eight to five boiling leaches and ultrasonic treatments were necessary to remove all the chloride from his samples, and suggested that some anion exchange may retain chloride ions on the clays. That possibility was also raised by Bischoff *et al.* (1970), and recent work on indurated clays also show some difficulties in chloride extraction by leaching (Moreau-Le Golvan, 1997). Experiments on marine sediments may not detect that effect, probably because of two main reasons: the high chloride content of the sediments and the pH range at which the leaching is conducted (Kriukov and Manheim, 1982).

Manheim (1974) claimed that the leaching technique, although possibly good for chloride, does not provide reliable results for cations. Those are more influenced, and often in a complex manner, by the type of sediment, dilution, temperature, duration of leaching, as well as dissolution or precipitation of phases. Besides, the extrapolation of leaching data to obtain the composition of interstitial solutions is a difficult matter, mainly because of the lack of behaviour linearity of the sediment/water system and because of the difficulties in achieving equilibrium in leaching solutions.

In 1985, Fontanive *et al.* tested the leaching method against the squeezing method on a number of Italian clays. They dispersed the sample in deionised water, with water/clay ratios ranging from 1.46 to 49, and separated the solution after a few minutes by filtration. By plotting the data, as suggested by Devine *et al.* (1973), they noticed that Ca^{2+}, Mg^{2+}, K^+ and HCO_3^- showed a non-linear behaviour (Figure 37). The extrapolation to the initial moisture content would provide an underestimation of the HCO_3^- content. A few tests conducted with longer equilibration times, up to one year, produced solutions with a higher dissolved salt content than squeezed waters. The conclusion of their study was that the method was time-consuming and unreliable, mainly because of the presence of readily-soluble salts in the natural clay samples.

Figure 37. **Washing-water compositions in relation to water/clay ratios: N = extrapolated porewater composition (after Fontanive *et al.*, 1985, 1993)**

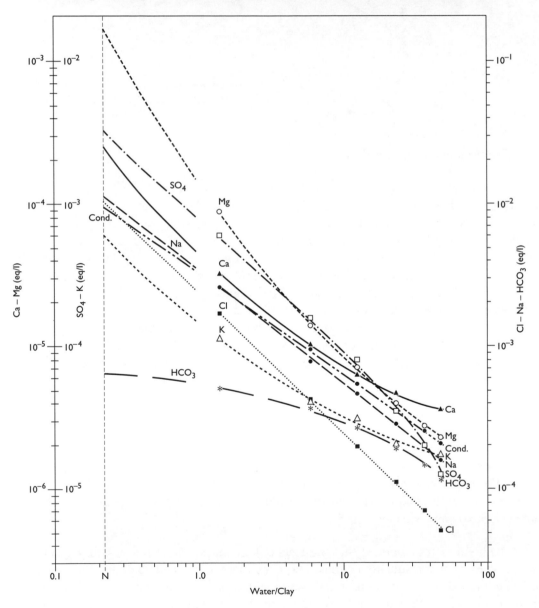

Iyer (1990) also reported a non-linearity of the dissolved species with increasing water contents, as well as higher concentrations in leachates with respect to solutions obtained by squeezing.

The studies of Fontanive *et al.* (1985; 1993) had pointed out the necessity of having reliable mineralogical data to interpret the leaching data. Von Damm and Johnson (1988) conducted an experiment reacting twelve shales with deionised water at 20° and 100°C. They checked both the solution chemistry and the mineralogical composition before and after the reaction with water. They used a fixed water/rock ratio of 1:4 and reaction times ranging from 38 to 43 d. Besides, the headspace of the bottles was purged with nitrogen prior to the reaction. Despite that precaution, oxygen may diffuse through the walls of the bottles. No attempt was made to control pH, resulting in very acidic solutions sometimes. They noticed that the sulphate ion was the major anion in every case, and sodium or calcium the major cations. They compared their results with the groundwaters found in the analysed formations. They concluded that sulphates were an experimental artefact resulting from the oxidation of pyrite, shales with no pyrite also showing the lowest sulphate contents of leachates. The mineralogical changes showed a loss of calcite and the formation of limonite at the higher temperature. The mineralogy was then compared with the minerals in equilibrium with the leachates using a geochemical modelling code. That helped them in trying to determine which minerals may be controlling the solution composition. They especially noticed that there was a strict relationship between the pH of the leachate and the occurrence of carbonates in the shales. pH in shales having carbonates, either observed in X-ray diffraction patterns or predicted thermodynamically, was close to neutrality. The authors concluded that the occurrence of pyrite was the most critical parameter in determining the solution acidity, while the presence of carbonates was critical to the buffering of the solution.

Oscarson and Dixon (1989) conducted a very similar experiment, checking both the solutions and the mineralogical composition of their clay samples. As they observed the presence of gypsum but not of pyrite in their samples, they postulated that the sulphate content of leachates derived from the dissolution of gypsum. On the other hand, the high sodium content would result from an uptake of calcium on the clay surface. They tried to recalculate the ionic strength of the pore solution, extrapolating the results obtained with different water/sediment ratios.

Those two studies helped in focusing the problem of the solid/water interactions during leaching. Both stressed the importance of an accurate determination of the mineralogy of the sample and the necessity of taking into account the dissolution of minerals and the exchange with clays. They also showed the necessity of controlling some parameters, such as the presence of free oxygen during leaching experiments.

Aquilina *et al.* (1994) reported the investigations of the Deep Geology of France Programme (*Géologie profonde de France*), in which aqueous leaching of core samples was conducted in order to constrain the data obtained from chemical monitoring of the drilling fluids (see Chapter II § 1.5.2). Freyssinet and Degranges (1989) presented the full details of the leaching experiments. They tested different parameters on some clay-rich materials, including the shaking time and method, the solid/liquid ratio and the sample hydration (leaching conducted on hydrated and dehydrated samples). They found that:

- No difference in concentrations was due to differences in the shaking method.

- Dehydration of the sample prior to leaching did not affect the composition of the calculated porewaters for a solid/liquid ratio of 1:2, but increased them for a solid/liquid ratio of 1:10.

– A decrease in the solid/liquid ratio would increase all the chemical elements, as would an increase in shaking time.

From those preliminary experiments, they concluded that the more "aggressive" the leaching, the more concentrated would be the leachate. Consequently, they selected a 4-h leaching time with a solid/liquid ratio of 1:2, hoping that those parameters would prevent "leaching due to cation-exchange problems". We have already discussed the basic problems involved in that approach, and we will not come back to them. Also, the chloride value in leachates is subsequently used to calculate the amount of rock leached by the drilling fluid at depth (see Chapter II § 1.5.2); that extrapolation is not valid if the kinetic evolution of concentrations during leaching is not well understood.

3.3.3   Modelling the solid/water interactions

Since it was shown that the chemical composition of the leachates may not be interpret without modelling the solid/water interactions, leaching has been used straightforwardly in some case studies only with regard to conservative elements, e.g. chlorine and bromine (Cave et al., 1994).

An attempt to reconstruct the porewater composition from data obtained by aqueous leaching using multivariate statistical deconvolution was reported by Cave et al., (1995) and Cave and Reeder (1995) on the previously considered data set. No specific information is provided on the mineralogy of the samples (Permo-Triassic red beds and Ordovician volcanic rocks). In their approach, they considered that solutes in leachates arise from a combination of sources, including residual solutes from porewaters, from contaminating drilling fluid, from fluid inclusions, as well as from the dissolution of minerals and non-linear water/rock interactions (e.g., ion exchange). If the leachate is derived predominantly from a linear combination of components and if non-linear interactions are minimal, the data may be modelled by principal-component analysis (PCA). They assumed that chloride and bromide are derived solely from the porewater component and that lithium, added as a tracer to the drilling fluid, allows the calculation of the drilling-fluid contamination. PCA is used to determine the number of components in a chemical data set derived from a number of different chemical sources. Besides, it allows the set-up of an abstract model of the components that may be used to test whether given components fit the data, according to a procedure known as "target factor analysis" (TFA). In the case study, the composition of the components was not known, but derived from the geological knowledge of the system and the estimation of some individual constituents of the test factors. Consequently, they first applied that deconvolution method on a simulated data set in order to verify the methodology, and secondly they used it on the real data set of aqueous leachates from the Sellafield borehole. As a result, the method applied to the real data set produced data that were consistent with independently derived groundwater data from the borehole for most determinants (hydrogen carbonate, calcium, barium, strontium, total inorganic carbon [TIC]). Potassium, silicon, sulphate, and aluminium showed instead values significantly different from the corrected values. As a conclusion, that method seems to have a potential application in a wide range of studies, but further work is needed to understand fully the limitations of the procedure.

That approach has been recently applied for the investigations on the Gard and Meuse sites (Reeder et al., 1997), the Tournemire site (Cave et al., 1997) and the Mont Terri project (Thury and Bossart, in preparation). It should be noted that, in all cases, the attribution of a given solution chemistry to a solution type (porewater, readily-soluble salts, contamination, etc.) is completely arbitrary and without any physical meaning. Besides, that approach completely neglects cation-exchange phenomena that are very important in clayey systems, as we have seen before. The use of very refined statistics does not provide reliable results if data are obtained using such a heavily disturbing technique.

Finally, in 1990, Bradbury *et al.* proposed to investigate the water chemistry, sorption and transport properties of marls using a leaching procedure. Their concept was applied to the Palfris marl (Baeyens and Bradbury, 1991; 1994) and to the Opalinus clay (Bradbury and Baeyens, 1997). In their approach, different leaching solutions are to be used, thus controlling alternatively a series of parameters (calcite saturation, pH, pCO_2, etc.). Basically, two types of ion extractions are conducted. The first is an aqueous extraction with different solid/liquid ratios, while the second uses nickelethylenediamine, a highly selective complex displacing all exchangeable cations from the clay minerals to the solution. The aim is to separate the contribution of soluble salts from the cation exchange on clays. The main objective of those studies is to calculate the composition of a synthetic solution, the laboratory standard marl solution (LSM) that would be in "chemical equilibrium" with the solid phase. That is a totally different approach, encompassing both the problem of water extraction and the quantitative definition of the abundance and characteristics of the "inner" and the "outer" solutions. The method relies extensively on geochemical modelling; as a result, it is only mentioned here, but it will be discussed in detail in Chapter III § 3.

3.4 Distillation technique for stable isotope analysis

The set of techniques described in this section uses the principle of distillation for extracting water from solid samples. Those techniques are consequently only to be used for isotopic measurements, since solutes are left behind in the matrix.

3.4.1 *Vacuum distillation*

3.4.1.1 *Historic development*

The vacuum-distillation technique has been set up since the beginning of isotopic measurements by mass spectrometry. It was first designed for extracting water from hydrous solid phases, such as volcanic glasses (Friedman and Smith, 1958), gypsum (Gonfiantini and Fontes, 1963) or phyllosilicates (Godfrey, 1962). The application of that technique to soil-water extraction was developed in the 1960s, when the IAEA initiated a project aiming to establish the relationship between the concentration of environmental isotopes (deuterium, oxygen-18 and tritium) in rain and infiltration water (Sauzay, 1974). Naturally, investigations turned on arid-zone hydrology (Allison *et al.*, 1983; Barnes and Allison, 1983; 1984; Aranyossy and Gaye, 1992) and consequently to low water-content systems. Recent developments of the studies for nuclear waste disposal sites focused on clay-rich environments with very low water contents. The design of the extraction did not undergo drastic changes, but the procedure parameters have been adapted to the different needs and interests.

3.4.1.2 *Set up and extraction procedures*

Basically, the solid sample is placed in a vacuum line (Figure 38) and heated: the released water molecules are trapped with liquid nitrogen. The water obtained is then used in conventional preparation lines for isotopic analysis.

In early studies, as the structural water was of interest, special care was devoted to the degassing of the sample in order to eliminate adsorbed water. That was achieved by preheating the sample at 100°C or more under vacuum. Then the extraction stage made use of a temperature according to selected minerals or materials, normally higher than 1,400°C for silicates. The evolution towards unsaturated-zone hydrology changed the focus from crystallisation water to porewater. The authors directly transposed the technique by selecting a lower extraction temperature. That may lead to

Figure 38. **Soil vacuum distillation line (after Jusserand, 1980) 1 = 10⁻³m Hg vacuum;**
2 = vacuum pump; 3 = traps; 4 = liquid nitrogen;
5 = 100°C heater; 6 = sample; 7 = thermostated bath

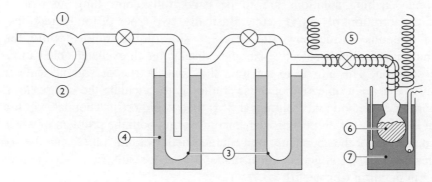

incomplete recovery of the water, and consequently, as we will see later, induce fractionation. Unfortunately, early work somehow masked the problem for a few years due to many reasons:

– Studies mainly focused on tritium and stable isotopes (Zimmermann *et al.*, 1966; 1967; Smith *et al.*, 1970), establishing depth profiles in soils. The tritium peak was easy to detect and samples needed electrolytic enrichment prior to analysis. The error is consequently much more difficult to show. In addition, the establishment of depth profiles using only one isotope may give consistent data, but not necessarily precise ones.

– The application to soils, where evaporation is one of the major natural process affecting the isotopic composition of water, did not allow to reveal artefacts due to the extraction and conditioning

In order to check for possible artefacts related to incomplete extraction, a simple calculation of the extraction yield may be used. Jusserand (1980) defined the extraction yield as:

$$R = \frac{m_1}{m_1 + m_2} * 100$$

where m_1 is the mass of water extracted by distillation and m_2, the loss of weight of the distilled sample when placed in an oven at 105°C ($m_1 + m_2$ being the total water content of the sample).

The temperature at which distillation is performed and the extraction yield are two essential parameters that should always be specified when describing the extraction procedure. While there is an international agreement on the need to control the extraction yield, each laboratory seems to select a distillation temperature that normally ranges from 50° to 110°C (Walker *et al.*, 1994). Many other parameters may vary according to different laboratories in the sampling, sample conditioning and water extraction procedures. Some case studies are reported in Table 8, to give an idea of their variability. Those data are reported also because, while a number of authors consider that the extraction yield is the only parameter affecting the isotopic composition of extracted solutions, others think that, especially in clay-rich media, the performance of the technique may also be affected by the sample pre-treatment and other factors.

France-Lanord and Sheppard (1992) and France-Lanord (1997) have performed distillation extractions at different temperature stages (0°-120°C, 120°-200°C and 200°-1,100°C). Those temperature stages have been selected in order to extract respectively the free porewater, the strongly

Table 8. **Parameters used in different studies for vacuum distillation**

Reference	Material	Temperature of extraction	Duration (hours)	Check for yield	Pre-treatment	Notes/Comments
Friedman and Smith, 1958	Volcanic glasses	1,450°C	1	Not specified	Sample degassing in the line at 110°C for 30-60 min	Traps with liquid N_2
Godfrey, 1962	Hydrous silicates	1,300-1,600°C	1-2	Not specified	Sample degassing at 458°C for 90 min (165°C for biotite)	Samples are crushed, sieved and minerals are separated with heavy liquids
Gonfiantini and Fontes, 1963	Gypsum	400°C	Not specified	Yes	Not specified	Gypsum crystallisation water
Smith et al., 1970	Soils	100°C (chalk) 110-150°C (clays)	24 12	Not specified	Samples are isolated from the atmosphere in the field	Tritium
Gouvea da Silva, 1980	Soils	70°C	Not specified	Yes	Samples are frozen to −80°C and pumped	
Saxena and Dressie, 1983	Soils	50-60°C	5	Yes	Samples are placed in the line with their cans	Traps at −60°C
Allison et al., 1987	Soils	50°C	Not specified	Not specified	Not specified	Traps with liquid N_2
Ingraham and Shadel, 1992	Soils	110°C	7	Yes	Dried and doped with known water	Traps with liquid N_2 Dynamic vacuum?
Araguas-Araguas et al., 1995	Soils	100°C	7	Yes	Brief degassing of the system	Study on partial recovery, different temperatures and times used resulting in incomplete extractions
Mathieu and Bariac, 1995	Soils	70°C	Not specified	Yes	Not specified	Incomplete extraction is corrected using the Rayleigh equation
Ricard, 1993 Moreau-Le Golvan, 1997	Clays	60°C	14	Yes	Samples are crushed, frozen to −80°C and pumped for degassing	
Pearson et al., in preparation	Clays	105°C	48	Yes	Samples are crushed and lightly pumped	
France-Lanord, 1997	Clays	0°-120°C 120°-200°C 200°-1,100°C	Variable		Samples are crushed, frozen to −180°C and pumped	Study on the isotopic composition of the different types of water

absorbed water and the crystallisation water. The isotopic analysis (especially deuterium) performed on each fraction show appreciable differences in composition. Nevertheless, due to the very low amount of the two last types of water, their contribution to the cumulated isotopic composition is almost negligible.

3.4.1.3 Microdistillation

Microdistillation with zinc is a variant of vacuum distillation, and does only provide deuterium values. A small soil sample is placed under vacuum in a side arm of a reaction vessel containing zinc. The zinc is heated to 450°C, while the soil is heated to a temperature comprised between 100 and 200°C. The water vapour extracted from the soil reacts readily with zinc to form hydrogen used for mass spectrometry (Turner and Gailitis, 1988).

3.4.1.4 Evidence of artefacts

Since the comparative study of Walker *et al.* (1994) (see Chapter II § 3.4.3) a big effort has been made to describe the possible artefacts induced by vacuum distillation in order to make the appropriate corrections due to their effects. In this section, incomplete distillation is not considered.

Ricard (1993) tested different grain sizes, distillation times and distillation temperatures on samples coming from the Tournemire formation.

Grain sizes smaller than 2 mm and comprised between 2 and 5 mm were considered. The samples with a water content of 3.1% displayed no fractionation as a function of the particle dimensions. Another test using bigger (1-2 cm) and smaller (mortar pestle) grain sizes showed more variable results, both in water and in isotope contents. No indication was given if the sample was conditioned in a controlled atmosphere.

Distillation times, ranging from 20 to 48 h, did not seem to affect the isotopic composition. Instead, temperature effects were observed. If a sample has been distilled at 60°C and is brought up to 100°C, the last extracted water droplets are heavily enriched in deuterium. That value does not seem to correspond to clay-crystallisation water that is usually much more depleted (Sheppard and Gilg, 1996). According to the author, it may represent water strongly bound to the clay. Nevertheless, that water has a negligible effect on the global isotopic composition of the sample, since it is only present in very low amounts.

Moreau-Le Golvan *et al.* (1997) also found on Tournemire samples a grain size effect, mostly influencing oxygen-18, but also deuterium to a lesser extent (Figure 39). The extracted water is enriched in oxygen-18 as the grain size decreases, with discrepancies reaching 3‰. For deuterium, discrepancies remain within the bounds of analytical uncertainty. An extensive discussion on the possible origin of that artefact (access to a different type of porewater, exchange with atmospheric carbon dioxide or oxygen) is conducted, but is not conclusive.

The results of those experiments show that the technique may be biased by a number of artefacts. It is extremely important, in order to obtain consistent results, that a maximum of parameters be controlled during the sample conditioning (especially the grain size and the contact time with the atmosphere). For the consistency of the results, the set of parameters should be the same for all samples. As we have already stressed in the sample conditioning (Chapter II § 2), the use of an inert atmosphere to prevent evaporation is not suitable because pure gases are anhydrous.

3.4.2 Azeotropic distillation

Figure 39. **Grain-size effect (after Moreau-Le Golvan *et al.*, 1997)**

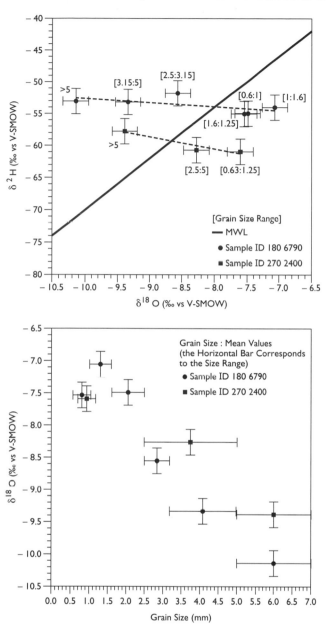

Azeotropic distillation was first used for extracting water from soils, adapting the method designed by Dewar and McDonald (1961). The technique is based on the observation that some solvents form an azeotropic mixture with water, showing a boiling point lower than the boiling points of the two end members. Different solvents have been used such as toluene (Allison and Hughes, 1983), xylene (Hendry, 1983) or petroleum ether (Shatkay and Magaritz, 1987). In the case of toluene (boiling point 110°C), the azeotropic mixture with water boils at 84.1°C.

Revesz and Woods (1990) carefully described the technique and the apparatus used (Figure 40). The soil was placed in a flask and covered with toluene. The temperature was raised to the boiling point of the azeotrope that evaporated, recondensing in the funnel with a cloudy appearance. At room temperature, the two liquids (water and toluene) separated. The temperature was gradually increased up to the boiling point of toluene. The latter condensed in the funnel with a clear appearance, indicating that the extraction was complete. The time required was usually much shorter that vacuum distillation, approximately 1.5 h. The collected water needed subsequently to be purified for traces of solvent. That was performed by adding wax to the solution and heating the mixture in a closed vessel. The wax dissolved the solvent and solidified on top of the sample at room temperature, allowing simultaneously to prevent evaporation from the vial during storage.

Figure 40. Azeotropic distillation apparatus for soil-water extraction (after Revesz and Woods, 1990)

3.4.3   *Evidence of artefacts*

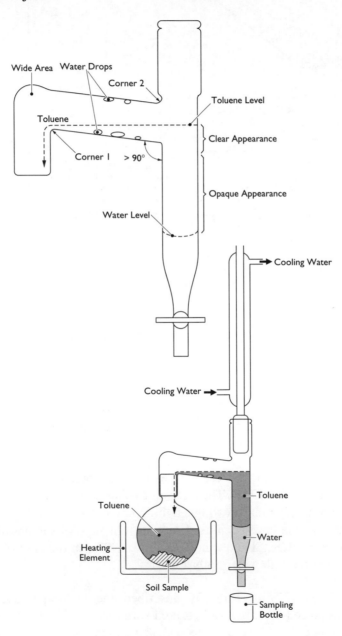

Revesz and Woods (1990) tested the technique for different soil types, variation in soil/toluene ratio, variation in grain size of the soil, dependence on the isotopic composition of the water, memory effects between successive samples, variability in water content and use of kerosene instead of toluene. No fractionation was observed, except in two cases:

- In gypsum-bearing soils, hydration water usually shows different isotopic compositions.[8] The authors suggest a simple method to check for the presence of sources of "water" *i.e.*, to compare the isotopic composition obtained by azeotropic distillation with the isotopic composition of a small amount of water added to the wet soil and extracted by decantation or centrifugation. If the two do not agree, then the azeotropic composition does not reflect the composition of the soil solution. If gypsum is present, the isotopic composition of the soil water must be obtained by isotopic-mass balance of the gypsum crystallisation water and the soil water.

- In soils with very low water contents, water is strongly bound to the particles and 100% yields are difficult to obtain. The authors tested clean sand with gravimetric water content of 3% and found that it was enriched compared to the conditioning water by +2.6‰ in deuterium and +0.56‰ in oxygen-18.

They concluded that the technique was simple, fast, inexpensive and suitable for any type of soil, providing good reproducible results (± 2‰ and ± 0.2‰ for deuterium and oxygen-18, respectively). Accuracy declines with low water contents.

Leaney *et al.* (1993) reported an important deuterium fractionation for soil samples previously conditioned. They chose two drying temperatures, 105° and 250°C, and after letting the samples cool down in sealed bags, they doped them with different amounts of isotopically-determined water (Figure 41). It appeared that fractionation was less important for samples dehydrated at 250°C. Besides, it increases as the clay content increased, reaching more than 10‰ in samples with 50% clay content. A comparison with porewater obtained by centrifugation again revealed a discrepancy of at least 3‰ along the depth profile. The authors consequently attributed a systematic depletion in deuterium of 3‰ to a fractionation of the azeotropic technique, while the rest of the discrepancy observed was attributed to the evaporation of the sample during the centrifugation extraction. It is interesting to notice that, in contrast with the results obtained by Revesz and Woods (1990), extracted water was systematically depleted with respect to doping water. Unfortunately, the study only concerns deuterium values and does not report the water/solid ratio used for the experiments.

3.4.3   Comparative studies: the effect of incomplete extraction

In 1994, Walker *et al.* launched an international comparison of methods to determine the stable-isotope composition of soil water. Four different types of soils (sand, gypseous sand, clay with low and high water contents) were prepared and distributed to 14 laboratories. Granulometry was less than 2 mm for all samples. With the exception of the gypseous sand, that was dried at room temperature in a dessiccator under vacuum, soils were oven-dried at 105°C. Bulk samples were allowed to equilibrate with atmospheric water vapour, and the average total "dry" gravimetric content was carefully determined. A predetermined water amount was then added to the solid samples in closed containers and shaken. Sample-to-sample variability was carefully tested using one extraction

---

8.   Gypsum ($CaSO_4*2H_2O$) contains crystallisation water with a different isotopic composition (Sofer, 1978). That water may be released at atmospheric pressure and in a hydrated environment at 42°C (D'Ans, 1968).

Figure 41. **Isotopic fractionation as a function of clay content (after Leaney *et al.*, 1993)**

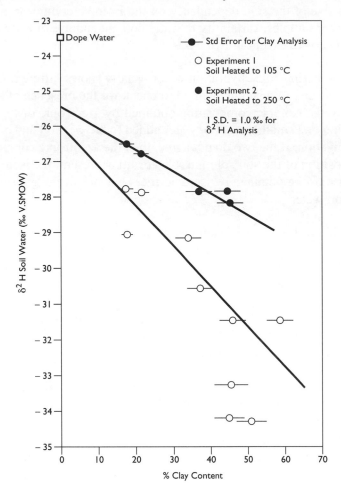

technique. Laboratories involved use a variety of water-extraction techniques, including azeotropic distillation, vacuum distillation, microdistillation with zinc and centrifugation.

The results obtained (Figure 42) show a large variation between laboratories in the isotopic composition of extracted water (up to 30‰ in deuterium and 3.4‰ in oxygen-18). The spread of results is most significant for the dry clay and the gypseous sand. Results are not randomly dispersed: they are systematically depleted with respect to the doping water and lie near a Rayleigh line for distillation at 35°C. In the case of gypseous sand, a shift towards the isotopic composition of gypsum crystallisation water is clear. Besides, a correlation with the final extraction temperature exists for all soils, especially for oxygen-18. The suggested cause of those variations is incomplete extraction, a phenomenon that tends to produce isotopically-depleted waters. The correlation with temperature suggests that heating (100-150°C) may be a necessary component for acceptable distillation results, especially when extracting water from clays.

That study highlighted the need to develop standard protocols for the extraction of water from soils. Besides, it started a serious effort aiming to understand the mechanism of distillation and to describe and model the influence of incomplete extraction. Gouvea da Silva (1980) had already noticed that incomplete extractions would lead to isotopically-depleted results. However, when she tried to model that depletion, she found that data would not follow the theoretic Rayleigh distillation process,

Figure 42. **Deuterium *versus* oxygen-18 plots for different soils. 1 and 2 denote the dope water value and a mass balance of dope plus residual water (from air-dried sample), respectively. Each point represents a participating laboratory (after Walker *et al.*, 1994)**

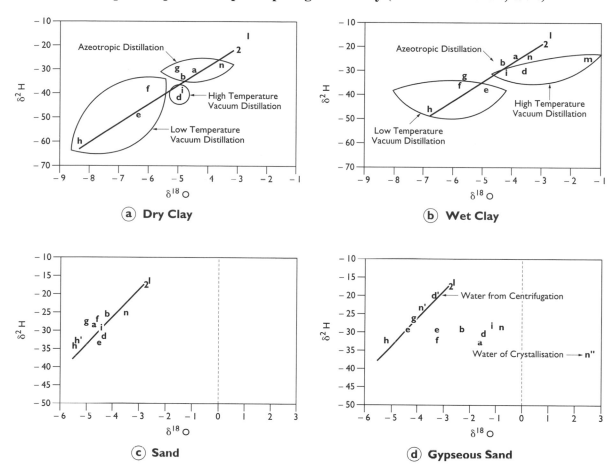

for the selected extraction temperature. Rayleigh equation needs the liquid phase to be in equilibrium at all times with the extracted vapour phase. If the extraction does not follow that equation, differences should be due to kinetics of the extraction procedure, possibly related, according to the author, to variations of the extraction temperature, the humidity of the sample or the kinetic difference between "free" and adsorbed water.

Ingraham and Shadel (1992) conducted experiments on the Hanford loam. The soil was oven-dried at 105°C for 24 h, then rehydrated to 4 and 8% weight with water having a known isotopic value. Samples were loaded in the extraction vessels and the soil water was extracted using either vacuum distillation (110°C for 7 h) or azeotropic distillation with toluene. The evolved soil water was collected during the extractions when a predetermined amount was released, providing six aliquots of water for each extraction. That allowed them to follow closely the evolution of the extracted water. By comparing those results with the Rayleigh distillation curve, it appeared that the observed evolution had the same trend as the theoretic one, but was shifted for both methods and especially in oxygen-18 (Figure 43). As a consequence, both methods seemed to fractionate the water sample during extraction, and neither method appeared to be accurate. The lack of accuracy was thought to be due to some infiltrated water being bound in the soil and not released below 110°C. The isotope effect of that "heat labile" water would be larger in low water-content soils as the bound water would represent a larger fraction of the total. Experiments conducted at very low water contents (3.6% and 5.2%) showed that extracted water was more enriched in oxygen-18 (up to 2‰) than doping water, while no effect was observed on deuterium. The change in oxygen-isotope composition was suggested to be the result of a hydration dependence of the oxygen isotope of soil water.

Figure 43. **Deuterium and oxygen-18 plots for (a) vacuum distillation and (b) azeotropic distillation (after Ingraham and Shadel, 1992)**

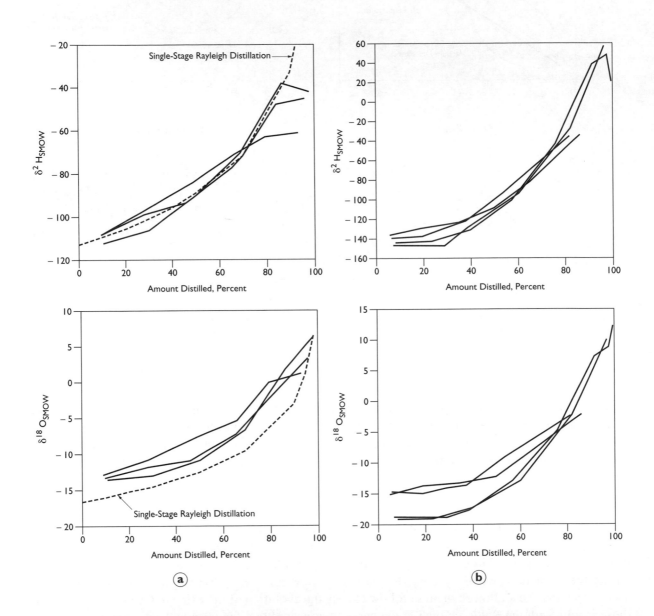

In order to test that hypothesis, a second set of experiments was designed, where soil was preconditioned with the same doping water, then dried and rehydrated. That way, the "heat labile water" in the soil should be in isotopic equilibrium with the test water, but not fractionated at its expense. Extractions performed with the two methods gave good results for the azeotropic distillation, while no improvement was observed for vacuum distillation. The authors came to the conclusion that, although neither method appeared to be accurate, azeotropic distillation should be preferred as only extracting "hydrologically active" soil water and not heat labile water.

Partisans of the vacuum distillation struck back in 1995. Araguas-Araguas *et al.* conducted an extensive study using the technique on different soil types with different clay contents. An oven-dried clean sand was doped with water and distilled with different duration times and temperatures, resulting in different extraction yields. Results showed that complete recovery of soil water may be obtained, with standard deviations for duplicate samples reaching about 1.3‰ and 0.14‰ for deuterium and oxygen-18, respectively. Those errors were only slightly higher than the analytical errors. Incomplete extraction (Figure 44) resulted in isotopic depletion: on the deuterium *versus* oxygen-18 plot, data fell around the meteoric water line, suggesting equilibrium conditions during the extraction of soil moisture. Using the Rayleigh equation, trends indicated that most of the extraction was obtained between 10° and 50°C, much lower temperatures than the ones used. The authors justified that as a delay in reaching the final distillation temperature, indicating that most of the water was extracted at the initial stage of the process. the authors concluded that vacuum distillation was reliable for sandy soils with high water contents, provided that the extraction was more than 98% of the original water.

Figure 44. Deuterium and oxygen-18 plots for soil water extracted from sand (after Araguas-Araguas *et al.*, 1995)

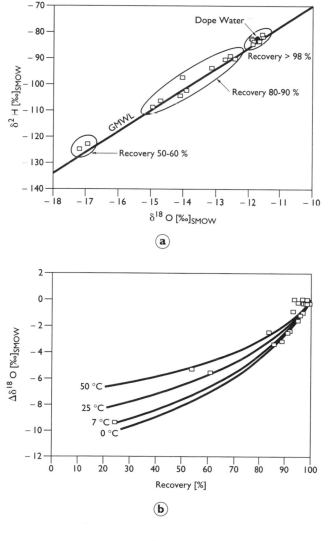

111

Percolation tests were conducted on a Brazilian soil (75% kaolinite, 80% clay content, 42-44% volume water content) and on an Austrian soil (50% clay including kaolinite, illite and chlorite, 34-34% volume water content). The innovative aspect of that approach was that the soils were not dried and conditioned, but flushed with the doping solution. The procedure avoided artefacts that may arise from drying and rewetting (see Chapter I). Soils were packed in columns and percolated with a known solution (Figure 45), the percolate being sampled at regular intervals (percolate I). The columns were then dismantled and part of the soil was distilled for 8 h at 100°C (soil water I). The rest of the soil was oven-dried at 70°C overnight to remove free water, then repacked in columns and percolated again (percolate II). Soils were distilled once more (soil water II) and redistilled at 350°C to remove all the water (high-temperature distillation). Results for the high clay-content soil indicated that the percolating dope water gradually replaced porewater, but, when extracted with vacuum distillation, the true isotopic composition of dope water could not be recovered. In the second percolation test, the significant isotopic shift would correspond to the water fractionation during the rehydration of the clay. On a deuterium *versus* oxygen-18 diagram, the data plot on a line that crosses the area of the water extracted at high temperatures. Those results indicated the presence of a pool of weakly bound water in the clay, which should be slightly depleted in both isotopes when compared to the free water (percolate). In the case of the Austrian soil, results did not show a clear trend as in the previous case, but formed a cluster depleted in oxygen-18 (0.26‰) and enriched in deuterium (3‰). In that case, the weakly-bound water in the soil should be enriched in oxygen-18 and depleted in deuterium with respect to free water.

The authors identified the reservoir of bound water in the hydration sphere of "cations of the clay particles" (interlayer water). That reservoir largely remained unaffected during the water extraction at 100°C. Bound water should be isotopically depleted in deuterium. That conclusion would be in agreement with the results obtained by Stewart (1972), who showed that water bound on kaolinitic mineral surfaces displayed approximately the same deuterium content than the vapour phase of bulk water in the reactor. In other words, the fractionation between free and bound water would be the same as the one existing between free water and its vapour phase in equilibrium. The observed shift in oxygen-18 would depend on the type of soil. The observation that vacuum extractions always provided results that shifted from the isotopic composition of dope water led the authors to suggest to use low extraction temperatures (< 100°C) for both vacuum and azeotropic distillations, if only free water was relevant.

Finally, Moreau-Le Golvan (1997) examined the effect of incomplete water extraction on Tournemire clay samples (Figure 46). Data did not follow the Rayleigh distillation curve for the selected temperature (60°C), or curves for lower temperatures. Besides, the regression line on the deuterium *versus* oxygen-18 plot indicated that the extraction proceeds out of equilibrium. Kinetic effects were taken into account using the model established by Gonfiantini (1986). In that case, the slope of the regression line agreed with the measured values, but the isotopic fractionation did not. The effect of isotopic fractionation was less than expected from both the Rayleigh and the Gonfiantini models. The author attributed that discrepancy to other factors that are not accounted for, namely the influence of water bonding on the kinetic fractionation factor and the diffusion coefficients used (established in air and not in vacuum).

Moreau-Le Golvan (1997) also pointed out that, in order to establish the extraction yield, the definition of the temperature at which all the water was released was of primary importance. The total water content is normally measured by oven-drying at 105°C for 24 h. In the case of the Tournemire argillites, thermogravimetric analysis showed a major weight loss before 100°C (2.6%), normally attributed to the extraction of free and adsorbed water. Nevertheless, between 100° and 400°C a smaller weight loss seemed to occur (1.39%). That weight loss might still be due to water loss, and in the case

Figure 45. **Results of the column experiment with the Brazilian soil (after Araguas-Araguas *et al.*, 1995)**

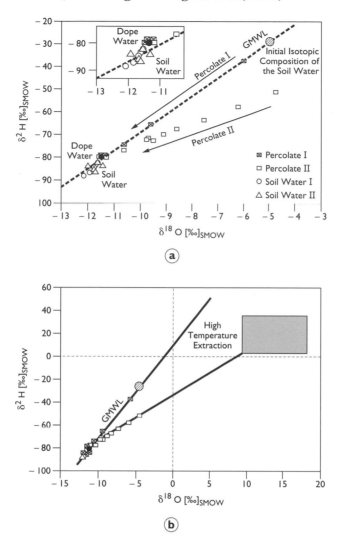

of very low water-content systems, it is of great importance for the definition of the total water content, and consequently the extraction yield.

3.5 Direct analysis of the isotopic content of water

In this section, we will consider techniques that do not involve the extraction of the aqueous phase from the solid sample, but the reaction with an external component leading, through some corrections, to the evaluation of the isotopic content of the water.

3.5.1 Equilibration with carbon dioxide (oxygen-18 analysis)

The conventional technique for oxygen-18 analysis of water samples involves an equilibration at 25°C with carbon dioxide, with a known isotopic composition. The isotopic exchange occurs through dissolution of the carbon dioxide in water. Equilibrium is known to be reached in a few hours (Epstein and Mayeda, 1953) according to the reaction:

Figure 46. **Effects of incomplete distillation (after Moreau-Le Golvan, 1997)**

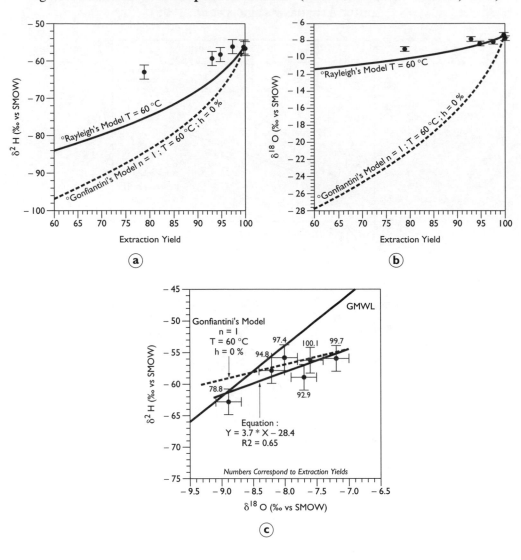

$$C^{16,16}O_2 + H_2\,^{18}O \leftrightarrow C^{16,18}O_2 + H_2\,^{16}O$$

Jusserand (1979) suggested that a direct equilibration of the solid sample with the carbon dioxide may provide the oxygen-18 values of porewater. For his experiments, he used five types of soils with variable clay contents (4-98%) and variable clay types (illite, chlorite, kaolinite and interstratified clays). The soil samples were previously dried at 105°C and rehydrated with different amounts of isotopically-known water. In addition, two water densities were selected (1.021 and 1.015). Water contents ranged from 9.26 to 33.80%. A known amount of isotopically-determined carbon dioxide was put in contact with the sediment for 7 d at 25°C, and subsequently extracted and analysed by conventional mass spectrometry.

Raw results needed a correction, based on the equation established by Craig (1957) for mass-spectrometry measurements adapted to the experimental conditions used:

$$\delta_c = 1.092\left(V_w^{-0.047}\right) \cdot \delta_m + 2.990\left(V_w^{-0.0033}\right)$$

114

where δ_c and δ_m are the corrected and measured deltas (in permil *versus* SMOW), and V_w, the volume of water used (*i.e.*, in that case, the volume of water injected in the solid sample). Consequently, the smaller the amount of water, the greater the correction. Corrected deltas were systematically more enriched (0.15 to 0.9‰) than the injected water: discrepancies were larger for low water contents and fine-grained sediments. Those samples also showed poor reproducibility. That behaviour was attributed not only to uncertainties related to the approximations used to define the parameters of the experiment, but also possibly to a fractionation effect existing for that type of clay-rich soils. Nevertheless, the technique of direct equilibration is considered reliable for samples showing relative humidity greater than 9%, with an uncertainty of ±0.5‰. No systematic effects were observed with changing water densities, and no tests have been performed with longer equilibration times.

Additional results were reported by Jusserand (1980) on natural and artificially-prepared samples. In that study, the technique was compared with centrifugation, vacuum distillation and low-pressure squeezing. Again, enriched values were reported and attributed, at least partially, to an overestimation of the applied correction.

Ricard (1993) tried the direct equilibration with carbon dioxide on some Tournemire claystone samples. Two tests were made. The first test was performed with a clay sample, dried at 100°C and rehydrated with isotopically-determined water. The sample was split in three subsamples, each of them analysed after different equilibration times (7, 18, 27 d). Results showed a depletion in oxygen-18 (up to five deltas) compared to the doping water used. On the other hand, it appeared that the solid samples had a higher water content than expected, possibly indicating that, after drying, the samples had adsorbed air moisture during the conditioning process. A second attempt made use of three different natural samples from Tournemire. Each was split in two, one part used for vacuum distillation and the second for direct equilibration. In that case, direct equilibration produced more enriched values, approaching within the error, after five months of equilibration time, the values obtained by vacuum distillation. It was not stated if any type of correction of the raw results had been applied, according to the Jusserand (1980) equations. From those results, the author considered direct equilibration as a valuable technique for the isotopic determination of porewater in that type of rocks, provided that the equilibration times were carefully checked. Nevertheless, the duration of the equilibration would make the use of that technique difficult on a routine basis.

3.5.2   *Direct equilibration with water (oxygen-18 and deuterium)*

That technique has been very recently developed at the University of Heidelberg and has only been presented at informal meetings or used in preliminary investigations (Rübel *et al.*, 1998). It relies on the isotopic equilibrium established in a closed container between the porewater of the sample and a known amount of isotopically well-defined water. Two experiments need to be run: one using a test water with an isotopic composition close to that of the porewater, the other with an isotopic composition significantly different. Sodium chloride is added to the test water samples (conc. 20 g/l) to avoid condensation on the walls of the container. The exchange is obtained through the vapour phase, by molecular diffusion. Equilibrium is reached within 10 d, and the test water is analysed with the conventional method of carbon-dioxide equilibration. The two values obtained allow to solve the two equations relating the isotopic composition and the volumes of porewater and test water. Results appear to provide a good estimate of the porewater isotopic value, with an uncertainty of 1.5‰ in δ^2H and 0.4‰ in $\delta^{18}O$, while the porewater content may be obtained within an error of 5%.

The technique has been applied within the framework of the investigations in the Lauenburger clay (Gorleben, Germany) and in the Opalinus clay (Mont Terri, Switzerland). In the latter case, isotopic results obtained were compared with the values of water flowing from boreholes and with porewater extracted with the vacuum-distillation method at 105°C for 48 h. Direct-equilibration data were in good agreement with borehole-water data and were heavily enriched with respect to vacuum-distillation data. The interpretation attributed the isotopic discrepancy of –7.0‰ in δ^2H and –2.13‰ in δ^{18}O to an artefact caused by the distillation process (Pearson *et al.*, in prep.).

3.6 Other techniques

In this section, we will consider isolated attempts of water and solute-extraction techniques that were not found very suitable.

3.6.1 *Microwave extraction*

An attempt of water extraction for isotopic analysis using a microwave oven has been performed on frozen samples within the framework of the ARCHIMEDE project. Very few details of the experience have been reported in the final European Community report (Griffault *et al.*, 1996). δ^{18}O values appeared to be at least 3‰ enriched with respect to water extracted by conventional techniques. According to the authors, that enrichment may be explained either by interactions with decomposing organic matter or with clay minerals, or by condensation of the atmospheric vapour contained in the microwave oven, despite the fact that the experiment was conducted under nitrogen flow. The lack of deuterium analysis does not allow to discriminate between the possible explanations. The authors claimed a recovery corresponding to the moisture content of the sample. Despite that, an evaporation of the sample prior to the treatment may also explain the isotopic enrichment (Pitsch, personal communication).

3.6.2 *Lyophilisation*

Two frozen samples of the ARCHIMEDE project were treated by lyophilisation. The collected water samples showed an oily consistency and a strong smell of hydrogen sulphide. The aspect of the water sample already showed that the experience was not carefully conducted, as lyophilisation should only provide pure-water samples. Oxyen-18 values were again heavily enriched for the small-diameter core sample, but only slightly enriched for the larger-core sample. Oxidation of pyrite and reaction with water (in absence of free oxygen, as in the case of lyophilisation) may occur, according to the following reactions:

$$FeS_2 + 2H_2O + e^- \rightarrow H_2S + Fe^{III}O(OH) + HS^-$$
$$HS^- + H_2O \rightarrow H_2S + OH^-$$

leading to an isotopic change in the water sample. Nevertheless, according to the mass balance, the complete oxidation of the pyrite would only affect 10% of the extracted water. Consequently, other water/rock interactions have to be considered in order to explain the isotopic shift. Again, the lack of deuterium data only allows speculating about possible explanations. Besides, the highest shift is obtained for the small-core sample, that may have been affected by a partial defreezing during the coring process. According to the authors, the isotopic shift would then be somehow related to the thermal history and the storing conditions of the samples, because of possible ice sublimation. Once more, a problem of sample conditioning seems to have occurred. Lyophilisation may then be considered again for isotopic measurements, provided that special care is taken for the cold chain not to be broken. Besides, for very low water-content samples, pyrite oxidation may increasingly affect the water isotopic composition (especially for deuterium, that has not been addressed in this report).

3.6.3 Carbon-13 and carbon-14 on-core extractions

Vacuum distillation was used to obtain carbon-13 and carbon-14 analysis from dissolved inorganic carbon in rock porewaters (Davidson *et al.*, 1995). Samples used consisted of variably-saturated tuffs, that were isolated upon removal from the core barrel by plastic sheeting and heat-sealable aluminium/plastic laminate, and then flushed with pure nitrogen in order to prevent atmospheric contamination. The samples were placed inside a cell connected to the cold traps, and heated at 150-180°C. Distillation lasted 4 to 15 h per sample. Results were compared with those obtained by core squeezing, according to the procedure described by Peters *et al.* (1992) and by vacuum acid leaching of the previously distilled cores.

Data suggested that the carbon recovery from the cores was far from complete. In fact, vacuum distillation lowered the porewater content, resulting in a precipitation of calcite in the matrix. About 15-60% of the inorganic carbon was recovered by distillation, the remainder precipitating in the pores as solid carbonate or being left behind as adsorbed carbon. If no primary calcite is present, the combination of gas extraction and acid leaching should allow reconstructing the porewater isotopic content. Carbon mass balance in the extracted porewater indicated DIC contents significantly higher than values obtained by core squeezing. That is apparently not due to incomplete distillation, since water contents were carefully checked. That suggests the presence of an additional reservoir of carbon dioxide in the rock, represented by adsorbed carbon.

Incomplete recovery results in an isotopic fractionation, favouring light isotopes in the extracted phase. That effect is too variable to allow corrections to be made for carbon-13, and attempts to reconstruct the original composition of porewaters were inconclusive. On the other hand, the impact of fractionation on carbon-14 measurements was significantly lower. The maximum fractionation computed using stable-isotope data, indicate a possible error within 4%. A comparison with data obtained with other techniques (squeezing and air measurements in the formation) were in good agreement. It should be noted that the samples used show quite a high carbon-14 activity (85-90 pMC), possibly masking the effects of an atmospheric contribution at some stages of the sampling and extraction procedure.

Moreau-Le Golvan (1997) set up a very similar vacuum distillation line for dissolved carbon extractions from the Tournemire argillites. In that case, the sample is previously crushed in a controlled atmosphere and is not heated. The extraction of substances other than pure carbon dioxide (kerogen?) interfered with the mass spectrometric measurements of stable-isotope contents. Carbon-14 results were significantly higher than background, indicating a possible contamination of the sample. Unfortunately, the impossibility to run an appropriate blank prevents the identification of the contamination source.

3.6.4 Radial diffusion method

Van der Kamp *et al.* (1996) described a method using a radial diffusion cell obtained from intact cores for determining the isotopic composition, the chemistry of porewaters and the effective porosity in aquitards. The diffusion cell consisted of a cylindrical sample of saturated porous material encased in a rigid, impermeable tube, with rigid impermeable seals at the top and at the bottom, to prevent water loss by evaporation and mechanical swelling of the sample. A hole drilled along the axis created a central reservoir, that was filled with a suitable test fluid (de-aired deionised water with appropriate tracers). The equilibration between the fluid in the central reservoir and the sample occurred by molecular diffusion: hence, successive episodes of water addition, equilibration and sampling, allowing the determination of the initial concentrations in the porewater and of the effective porosity of the material.

The method has only been tested in cores of Canadian clay-rich tills for chloride and sulphate ions, as well as deuterium, that are believed to be conservative and non-reactive. Water was injected five times in the central reservoir, and each injection included a waiting period until equilibration was reached. Electric-conductivity measurements show that the process was completed within 56 d. A careful measurement of the amount and composition of the water injected and withdrawn allowed to plot the concentration of solute removed to the net mass added to the solid sample. In fact, it was found that, for deuterium, the net mass added was positive, being the initial solution artificially enriched in heavy isotopes. Chlorine and sulphate ions diffused instead from the solid to the water, and the net mass added was thus negative. A linear regression, on the plots of concentration *versus* net mass added, provided values for the original porewater concentration directly as "y intercept".

A comparison of those values with the groundwater composition obtained from boreholes showed a good agreement for deuterium, within the range of analytical error. Chloride values were systematically higher (approximately 30%): the authors explained that the discrepancy was due to a contamination of the diffusion cell. Sulphate contents were also slightly different, but do not show any systematic variation. That indicates that, if they were influenced by oxidation and reduction phenomena, those effects were not very significant.

In our opinion, despite the very promising title of the paper, the technique seems valid only for obtaining deuterium values and possibly chloride contents, but it should not be used for deriving porewater compositions. In fact, it only considers diffusion from and to the pores, neglecting salt dissolution and exchange/adsorption phenomena. In that respect, the approach is biased by the same conceptual errors of deionised-water leaching, as shown by the results obtained even for the most conservative ions. The performance of the technique for obtaining deuterium values should be tested on other types of clay-rich materials, since it assumes no swelling in the inside hole and consequently no change in the pore size and access.

In a companion paper, the concentration change in the reservoir during equilibration, adequately modelled using Laplace transform method for both finite and semi-infinite domains, was used to derive the effective diffusion coefficients, as well as adsorption and effective porosity (Novakowski and van der Kamp, 1996).

3.7 Organic-matter extraction

Mostly soil scientists and petroleum geologists have designed procedures for extracting organic matter from clay-rich samples. Those procedures vary according to the type of relevant organic matter: the water-soluble fraction, the solvent-extractable fraction or the non-extractable fraction.

As we have seen in Chapter I, water-soluble organic matter is mostly made of humic substances. Investigations have concentrated on the extraction, purification and characterisation of those substances, mainly because they are the easiest to process. In addition, humic substances are most likely present in clay porewaters and are considered responsible for the complexation and transport of some radionuclides.

Despite the efforts made, there is still a lack of agreement on the structure of humic substances, that reflects the complex processes responsible of their formation (Schnitzer, 1978; Sposito, 1984). The aim of the extraction procedure is to break the bonds between the humic substances and the solid particles without changing the structure and the functional groups of the organic matter itself, in order to use the extracted substances for speciation studies. According to Hayes (1985), water is an excellent solvent for humic substances when the acid groups on the macromolecules are dissociated and their structures expand to allow the solvation of the polar moieties in the structures. The dissociation is best accomplished with the use of strong basic aqueous solutions, but those may to some extent degrade the macromolecules and lead to the formation of artefacts. When the charges on the humic molecules are neutralised by polyvalent cations, dissolution in water may be inhibited, because cations tend to condense the structures by forming short bridges between charges on adjacent strands of the macromolecule. In that case, the use of neutral-salt solutions that complex and remove polyvalent cations may help in the extraction of humic substances.

A standard procedure has not been designed yet, and an international study group (the International Humic Substances Society) is currently working on it. The most commonly used procedure is summarised in Figure 47 and involves washing the sediment with a mild alkaline solution (*e.g.*, 0.1 M of sodium hydroxide) in order to dissolve humic substances. The insoluble fraction, including the inorganic fraction, the non-humic substances and humin, is separated by centrifugation and discarded. The solution is then acidified to a pH of 2 (*e.g.*, with hydrochloric acid), to separate precipitating humic acids from fulvic acids, that is, the low molecular fraction that is soluble at any pH.

Figure 47. **Extraction procedure adopted for the separation and purification of humic substances (modified, after Testini and Gessa, 1989)**

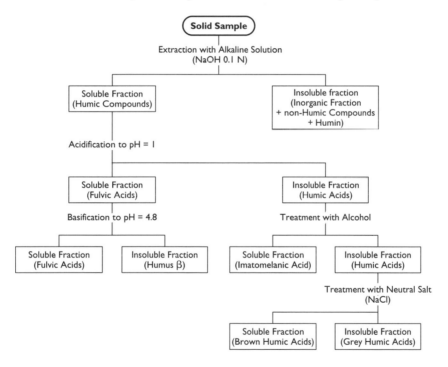

Humic acid may be further treated with alcohol to dissolve imatomelanic acid. The insoluble humic acids still may be separated into brown humic acids (soluble) and grey humic acids (insoluble) by processing them with a neutral salt like sodium chloride. Fulvic acid may be purified by raising the pH to 4.8 to separate a component called "humus β". The purification is usually obtained by repeating the acid-base treatment several times and combining the different fractions, or by performing ultrafiltration at selected molecular cut-offs. Characterisation of the extracted compounds is performed through elemental analysis (carbon/hydrogen/nitrogen ratios), ultraviolet-visible spectroscopy and IRS, NMR and electron-spin resonance (ESR).

The first investigations on the Boom clay (Beaufays *et al.*, 1994) focused on the isolation, purification and characterisation of humic and fulvic acids in order to test their complexation properties with respect to europium (taken as analogue to americium) and the effects of radiolysis on complexation. The organic matter was extracted both under oxidising and reducing conditions. The extraction procedure was initiated by suspending the clay in a carbonate buffer (1:1 ratio), consisting in a chemical analogue of the Boom-clay interstitial solution. The slurries, equilibrated for 2 d, were separated by centrifugation, and the supernatants containing humic acids were collected. The procedure was repeated three times, and the extraction yield ranged between 4 and 10% of the total organic-matter pool. The reduced organic matter was further purified by high-speed centrifugation and kept under nitrogen atmosphere. If necessary, excess salt would be eliminated by dialysis or ultrafiltration with a selected molecular cut-off range. The oxidised organic matter was more drastically purified by lowering the pH to 1 with nitric acid, leading to a quantitative coagulation of humic acids. The supernatant containing fulvic acids was decanted, and the humic acid fraction redispersed in a 0.1 N HNO_3. That procedure was repeated, normally twice, until the supernatant was clear, thus confirming the absence of fulvics. Humic acids were finally separated by centrifugation, followed by redispersion in distilled water, then dialysed until the pH reached approximately 5.6. The suspension was then stored in the dark at 1-2°C. According to the authors, those rather mild working conditions should ensure that only minor changes are caused to the organic matter and guarantee the representativity of the *in-situ* conditions. The characterisation that followed made use of potentiometric titrations, ultraviolet-visible spectrophotometry and CEC measurements (by saturation with trivalent cobaltihexammine labelled with cobalt-60).

In the framework of the ARCHIMEDE project (Griffault *et al.*, 1996), the extraction of the soluble fraction of the organic matter has been performed using a 0.2‰ solution of an alkaline detergent (RBS™), at a pH close to neutrality and a clay/water ratio of 1:5. All operations were conducted in an inert atmosphere. The solution was shaken for 13 d at 20°C and separated in a centrifuge. A second high-speed centrifugation stage allowed the separation of a black liquid from a black solid. Since no characterisation was conducted onto those two organic fractions, their representativity is uncertain. The extraction yield was 30 times greater than the extraction with deionised water, but no calculation on the extraction yield are reported. Cation-exchange capacities measured with cobaltihexammine chloride on the black-solid residue showed to be approximately double the values obtained in the previous study (Beaufays *et al.*, 1994). That discrepancy clearly indicates that different fractions of extractable organic matter have been separated according to the adopted extraction procedure, and that more effort has to be put in the formulation of a standard procedure to allow the intercomparison of results.

More recently, repeated extraction cycles with acidic (pH = 2) and basic (pH = 12) solutions provided a high recovery yield of organic matter from Boom clay, with the separation of the water-soluble and non-extractable fraction (Dierckx *et al.*, 1996; Devol-Brown *et al.*, 1998).

The water-insoluble fraction of the organic matter (bitumen) may be obtained using organic solvents, such as heptane, cyclohexane, toluene or chloroform (Griffault *et al.*, 1996). Extracted products may be identified by gas chromatography and mass spectrometry. Finally, the non-extractable organic fraction may only be obtained by destruction of the inorganic particles by strong acid attack. That fraction may be characterised to provide information on the origin of the organic matter (marine or continental), its maturity, the thermal history and the petroleum potential, and ultimately, some indication on the possibility of fluid migration (oil and water) occurring in the formation. Elemental analysis indicates which type of degradation affected the organic-matter pool and which organisms were responsible for it.

A complete organic matter characterisation, including petrological, bulk organic geochemical, aromatic and aliphatic biomarkers and light hydrocarbon (LHC) has been conducted in the framework of the Mont Terri project. LHC-distribution analysis may shed some light on the migration and subsurface redistribution of those very mobile compounds. LHC cannot be extracted using solvents, but need to be released by a mild thermal treatment in a headspace apparatus or by gas stripping. That characterisation allowed to differentiate three main sources of organic matter in Opalinus clay (Pearson *et al.*, in preparation). Due to their very low quantity, humic substances could not be extracted from borehole-water or squeezed-water samples. Nevertheless, they are believed to constitute the bulk of the total organic carbon (TOC).

PART III

PROCESSES AND CURRENT INTERPRETATIONS

1.   EXTRACTION TECHNIQUES

In the previous chapter, we examined the available techniques for extracting water from clay-rich sediments and rocks, discussing the possible artefacts that have been evidenced for each technique. This section aims to bring an insight on the processes that are likely to occur when disturbing the clay/water interface by subtracting or adding solvents, applying or eliminating pressure, etc.

This third chapter of the report is based on the available comparative studies, that use different techniques on samples with controlled mineralogy and water content. Studies of that type are very rare, and synthesis mainly concern applications to soils (Adams *et al.*, 1980; Jusserand, 1980; Litaor, 1988; Iyer, 1990; de Lange *et al.*, 1992; Walker *et al.*, 1994; Angelidis, 1997). The review is also based on a number of internal reports concerning investigations conducted on low water-content clay formations (Gard, Meuse, Mol, Mont Terri, Tournemire) for the definition of their geochemical characteristics (Bath *et al.*, 1989a; 1989b; Blackwell *et al.*, 1995a; 1995b; Brightman *et al.*, 1985; Cave *et al.*, 1997; De Windt *et al.*, 1998b; Griffault *et al.*, 1996; Reeder *et al.*, 1992; 1993; 1997; Thury and Bossart, in preparation). In some of those cases, the same technique has been applied to obtain solutions from quite different types of clays (see Chapter I § 5 for the site description).

This section focus on processes occurring at the solid/water interface and relies on the review of the fundamental properties of the clay/water interactions that was made in Chapter I, in order to compile a list of possible artefacts related to extraction techniques. Firstly, processes that may be demonstrated in experimental studies are described. Secondly, the attempts to correct the artefact effects that were found in the literature are reported. Finally unsolved questions and problems are highlighted, stressing the areas where further investigation is needed.

1.1   Performances of the extraction techniques

A first indication on the performance of a given technique to extract water from a given sample is based on the force used to break the bonds between the water molecules and the solid. In other words, we should examine the problem from the energetic point of view, without considering, for the moment, the solute/clay interactions.

In Chapter I, we defined the matrix, osmotic and water potentials. We have also seen that clays show different matrix-potential curves, according to the surface charge they develop and to the cation population on the adsorption sites (Figures 5 and 11). Spectroscopic studies have demonstrated that, beyond 11 layers of water molecules, the properties of water resemble those of bulk water, while, closer to the surface of the clay particles, properties may vary significantly (Swartzen-Allen and Matijevic,

1974; Sposito and Prost, 1982). According to Van Olphen (1965), five layers of water on montmorillonite show a suction potential corresponding to a pF of 4.7, two layers to a pF of 6.4, and one layer to a pF of 6.7. The air-inlet point, corresponding to the beginning of desaturation for a clay like the Boom clay, for example, is situated at a pF of approximately 4 (Horseman et al., 1996). Logically, that value should be higher for indurated clays with lower water content.

A simple approximate calculation allows to determine the suction potentials for each extraction technique (Table 9).

Table 9. **Suctions applied by different water-extraction techniques**

Technique	Specifications	Applied suction		Reference
		MPa	*pF*	
Oven-drying at 105°C		1,000	7	Studer, 1961
Low-pressure squeezing	4 bars		3.6	Jusserand, 1980
Squeezing[9]		70	< 5.8	
		110	< 6	
		552	< 6.7	
Low-speed centrifugation	2,500 rpm for 20 min		3	Jusserand, 1980
Centrifugation[10]	7,500 rpm	1.8	4.3	Edmunds and Bath, 1976
	20,000 rpm with immiscible displacent	2.3	4.4	
Vacuum distillation[11]	complete extraction		7	Studer, 1961

According to those values, apart from low-pressure squeezing and centrifugation, all the techniques should be able to extract free water from clays, and most of the techniques are likely to affect also any water strongly bound to the clay surfaces.

The possibility of extracting not only free water, but also to some extent, strongly bound water is an important issue, especially for isotopic analysis. Indications on the behaviour of clay samples with respect to water extraction may be obtained by adsorption-desorption isotherms (Decarreau, 1990). It is normally observed that those show hysteresis phenomena, most likely because of the changes in the clay structure due to desaturation (layer collapsing, grain-size modifications, deswelling) (Boek et al., 1995).

---

9. Those suction values are calculated assuming that the applied squeezing pressure tranfers directly and completely to the porewater. They consequently represent a maximum value.

10. Those values are calculated using the equation reported by Edmunds and Bath, 1976, for describing the suction applied at the midpoint of the centrifuged sample (see Chapter III § 3.1).

11. That suction value is assumed, since the complete water extraction is checked by oven-drying the sample at 105°C.

Nevertheless, useful information on the amount of different types of water may be derived together with suction parameters, that should be considered only to extract free water. That type of analysis is not routinely applied in the investigations on indurated clays.

1.2 Processes related to water and solute extraction

1.2.1 Effect of increasing pressure

Squeezing and, to some extent, centrifugation involve a pressure increase in the interstitial solutions caused by the applied external force (piston or centrifuge). That may influence the chemical and isotopic composition of extruded solutions.

The effect of that pressure change has been extensively studied for high water-content, high salinity-systems (marine sediments). For that type of samples, the increased pressure effect is considered to be small or undetectable for pressures below 600 kg cm^{-2} (59 MPa) (Manheim, 1966) or 1 kbar (100 MPa) (Manheim and Sayles, 1974; Manheim, 1976). For higher pressures, both concentration decreases and increases are reported (Kharaka and Berry, 1973; see the review in Kriukov and Manheim, 1982). Those pressure limitations seem to apply also to high water-content, low-salinity environments (lake sediments) (Patterson et al., 1978). In all those cases, the ion concentration in the extruded pore fluids is constant for most of the squeezing sequence (down to a water content of 50% per dry weight), and the electrolyte concentration is close to the in-situ value as measured with selective electrodes (Kriukov and Manheim, 1982).

On the other hand, in case studies on low water-content and clay-rich rocks, high squeezing pressures need to be applied to drive water out. Important changes in the porewater chemistry have consequently been detected (Brightman et al., 1985; Reeder et al., 1992; Böttcher et al., 1997).

Two additional phenomena may take place simultaneously: oxygen intrusion in the pore spaces leading to a change in redox conditions, and a temperature-induced modification of mineral solubilities. Recent squeezing devices are equipped with a temperature-control system and the extraction is performed in anaerobic conditions (Reeder et al., 1998). Nevertheless, the solution composition still seems to change with the increasing stress, indicating that water/rock interactions are still occurring.

Only in a few studies on clays was water actually collected at different stages of increased pressure (Brightman et al., 1985; Rodvang, 1987), mainly because the low water content does not allow that type of investigation. Results normally show a decrease in concentration of major ions; in addition, small monovalent ions are more affected than large divalent ions. Reeder et al. (1992) squeezed, both aerobically and anaerobically, some Boom-clay samples. All elements (major and trace) displayed a sharp decrease in concentrations, except for magnesium, nitrate, bromide, TIC, TOC and silicon, that show anomalous tendencies and are first depleted, then enriched. Unfortunately, those results were not compared with those obtained by other techniques in the final report of the ARCHIMEDE project (Griffault et al., 1996). Nevertheless, Beaucaire et al. (in press) showed that squeezed samples, although having chloride levels comparable with those obtained in piezometers, have lower bromide content, concluding that the observed fractionation must be related to the ionic radius of the considered species.

Porewater extrusion by increasing pressure is normally considered analogous to the natural process of compaction occurring in sedimentary basins (Reeder et al., 1998). Nevertheless, the extrapolation of the natural system to the artificial one is not straightforward, mainly because squeezing pressures are relatively high and applied in a short period of time, not allowing water to move within the pores in the same fashion (Magara, 1976). When expanded layers are compressed, the total cation exchange capacity of the clay formation decreases. With compaction pressures up to 20,000 psi (138 MPa), divalent cations are released to a greater extent than monovalent cations (Weaver and Beck, 1971). Other studies on compacted clays, showed that the compaction pressure has no influence on the sorption thermodynamics of the material (J. Ly, CEA, personal communication)

Studies on compacted clay layers have shown, under laboratory conditions, the existence of an ion-filtration effect (Kharaka and Berry, 1973; Kharaka and Smalley, 1976; Graf, 1982). According to those studies, the anionic and cationic concentrations of extruded solutions are lower than in input solutions. The following retention sequence has been established from the most retained to the least retained:

$$Cs^+ > Rb^+ > K^+ > NH_3^+ > Na^+ > Li^+ > Ba^{2+} > Sr^{2+} > Ca^{2+} > Mg^{2+}$$

While some authors consider that phenomenon to occur during squeezing at high pressures (Rodvang, 1987), Kriukov and Manheim (1982) stated that it does not occur because the residual porewater shows an even lower electrolyte concentration than the extracted porewater. Unfortunately, the latter do not specify the method used to extract the residual porewater.

In other studies, the Donnan principle causing anion exclusion is considered responsible for the relatively high salinity of the first extracted solution drops (Kriukov and Manheim, 1982; Brightman et al., 1985). According to that principle, free porewater should have a higher dissolved salt content than water close to the clay surfaces, because in that region, the amount of ions is imposed by the surface charge of the clay particle. When the clay particles are compressed, their electric-influence regions overlap. That especially affects anions, since they are excluded from that double layer region. As a consequence, anions will be extruded first and will carry a sufficient amount of cations with them to preserve the electric neutrality of the solution. As the porewater is eliminated, the water in overlapping double layers would progressively dilute it.

According to that explanation, chloride obtained by squeezing and by leaching should display comparable values. In fact, it is assumed that squeezing only affects free porewater, while leaching affects both pore and interlayer water, but may be recalculated back to free porewater, if the moisture content of the sample is known. Chloride is assumed to be only present in the free porewater. Nevertheless, results are not directly comparable for indurated clays (Blackwell et al., 1995a, 1995b) and chloride contents are higher for squeezed samples than for leached samples. It should be noted that leaching data normally display higher ionic content than squeezing data, because of the dissolution of soluble salts. The reason for that discrepancy is not clear, but may be explained by the difficulty to back-calculate leaching data to the free porewater composition, or by an increased salinity of the water obtained by squeezing, or by a positive anion sorption on clays during leaching. That anomalous behaviour of chloride (and to a minor extent bromide) is not at present explained and may be related to a change in the alkalinity to keep the ionic balance.

Reeder et al. (1992) and Pearson et al., (in preparation) showed that an increase of the external pressure leads to a loss of dissolved gases.

Another possible explanation, that is rarely considered to justify the salinity changes of the squeezed solution, is a modified solubility of solid phases due to compression. Two contrasting forces particularly affect the carbonate system. On one hand, carbon-dioxide degassing of the porewater solution may lead to calcite precipitation (Pearson *et al.*, 1978); on the other hand, the pressure increase should thermodynamically cause calcite dissolution. The net effect of those two driving forces will most likely depend on the local (within the sample) conditions (Fritz and Eady, 1985).

There is a great interest in the evaluation of the effects of pressure differences on water chemistries. Preliminary results on artificial systems (compacted MX80 bentonite and experimental solutions) indicate that the concentrations of the squeezed porewaters are clearly lower than the equilibrated "external" solutions and tend to decrease with increasing density during squeezing (Muurinen and Lehikoinen, 1998). Ongoing research at CIEMAT (Villar *et al.*, 1997; AITEMIN *et al.*, 1998) within the framework of the European Community's FEBEX project, is dealing with changes in solution chemistry with the applied pressure. The clay sample in the squeezing cell is allowed to stand for one week after each step of increased stress. It was observed in some cases that water starts to outflow only a few days after the pressure increase due to the slow movement of the water in the pores (A.M. Fernández, CIEMAT, personal communication.). Analysis of the solution chemistry after each stage will be performed. In addition, the squeezed core will be dissected and analysed for moisture and dissolved-salt distribution, as well as for changes in the mineralogy. A detailed study on the topic is also foreseen among the investigations at the Mont Terri site (Thury and Bossart, in preparation).

Recent work on highly-indurated clays (Reeder *et al.*, 1997; Cave *et al.*, 1997) used a mixed technique to extract water, consisting in a partial rehydration of the sample prior to squeezing. Reeder *et al.* (1997) concluded that, for most elements, the average concentration found in the rehydrated squeezed water was in reasonable agreement with that found for non-treated samples. Exceptions to that were chloride, potassium and pH. Chloride was found to be depleted in rehydrated samples in an amount proportional to the ratio of the moisture content of the rehydrated sample to the natural moisture of the sample. Potassium was found to be enhanced in rehydrated cores, and pH increased, on average, by 0.5 unit. In our opinion, the results obtained by using such combined techniques should be cautiously considered, since the artefacts induced by both techniques are still poorly controlled.

Very few studies (Brightman *et al.*, 1985; Reeder *et al.*, 1992; Beaucaire *et al.*, in press) also examine the effects of an increase of the external pressure on the isotopic composition of the extruded solutions. Theoretically, a small membrane effect should be expected (Coplen and Hanshaw, 1973; Yeh, 1980; Benzel and Graf, 1984). The variability of the values seems more important for oxygen-18 than for deuterium, but the small amount of available data does not allow to draw any conclusion on the topic.

1.2.2   Effect of decreasing pressure

The decrease in external pressure and its consequences has been extensively studied. It is usually observed that the lack of confining pressure, such as the one encountered in the formation (lithostatic pressure) induces a decompaction of the rock sample materialised by the formation of microfractures, through which gas migration is increased. An immediate observed effect is the degassing of the sample (Zabowski and Sletten, 1991), together with the input of atmospheric gases. Incoming oxygen causes oxidation reactions to increase, and its effect will be addressed in the next section.

Degassing may occur whenever the gas partial pressures in the porewater are higher than the surrounding ones. The use of a controlled atmosphere will consequently not solve the problem, unless displaying the same gas partial pressures. Degassing has been detected during sampling, transport and centrifugation of soil samples (Dahlgren *et al.*, 1997), despite of all the efforts made to minimise carbon-dioxide loss with careful sampling in closed, centrifuge tubes. Between 86 and 99% of the total carbonate concentration were lost, resulting in a reduction of bicarbonate concentration up to 0.67 mM in 24 h. That phenomenon does not seem to be restricted to that extraction technique, and is also observed during squeezing.

Carbon-dioxide degassing is a relatively serious problem since it affects the carbonate system. Carbon-dioxide escape will lead to the loss of total dissolved carbonates, the precipitation of carbonates in the pore spaces, the subtraction of calcium from the solution and ultimately the increase of cation exchanges with clays. Calcite precipitation depends on the presence of calcium-carbonate crystallisation nuclei, and is consequently reported to occur mainly in carbonate-rich sediments (de Lange et al., 1992). Manganese is very sensitive to carbonate precipitation: negative "spikes" in the manganese content of porewaters from marine sediments coinciding with intervals of enhanced calcium carbonate content have been attributed to a decompressional effect (de Lange, 1986). Carbonate precipitation may lower the effective porosity and help trapping the solutions inside the core sample.

That degassing phenomenon is very rapid and is thought to affect also water sampling from boreholes (Pearson et al., 1978). In addition, as long as water is in contact with the clay, the pH will be buffered by water/rock interaction, but as soon as it is separated, the pH will start rising (Pitsch et al., 1995b; Beaucaire et al., in press). As a consequence, even pH measurements in the field are not reliable.

Carbon-dioxide loss alone has no effect on total alkalinity, but decreases the TIC, while calcite precipitation affects both parameters (Stumm and Morgan, 1996). Some authors overcame the problem of carbon-dioxide degassing by performing total alkalinity analysis immediately after porewater collection (Berner et al., 1970; Patterson et al., 1978; Emerson et al., 1980).

Attempts to evaluate and correct water chemistries and carbon isotopic composition for the effects of carbon-dioxide degassing have been made, based on thermodynamic modelling. Pearson et al. (1978), using carbon-dioxide partial-pressure field measurements, back-calculated an estimate of field pH and carbon isotopic composition of water samples before degassing. Two methods were described, using computer codes and isotopic-fractionation factors and assuming that degassing occurs at equilibrium. The effects of carbon dioxide degassing or ingassing on pH and calcite saturation indices are shown in Figure 48 for the calcite/carbon-dioxide/water system at 25°C.

For a given total alkalinity and initial calcite saturation index, the diagram provides the change in pH and the saturation index resulting from the carbon-dioxide addition or loss. Because alkalinity is independent of pCO_2, carbon-dioxide ingassing and degassing follow paths at constant alkalinity (solid lines). As an example, solution A (2.27 meq/kg alkalinity and –0.15 calcite-saturation index) loosing 4% of carbon dioxide will result in solution B, supersaturated in calcite (S.I. = +0.2) and with a higher pH (0.35 unit) than the original pH. Although that diagram only applies straightforwardly to simple calcium-bicarbonate water, it gives indications on how to proceed to evaluate carbon-dioxide-degassing artefacts.

Degassing also affects noble-gas measurements, despite the great care taken in conditioning the rocks as soon as they are cored (see Chapter II § 2). Neon, krypton and xenon, having an atmospheric origin, are less affected by degassing due to their higher solubilities in the porewater and lower diffusion coefficients. Helium is estimated to be lost in a proportion equal or less than 30% during handling (Pearson et al., in preparation). It is not stated if that degassing may influence helium's isotopic ratios as well.

Figure 48. **Summary of pH uncertainties and calcite-saturation index in calcium-bicarbonate waters open to carbon-dioxide gas at 25°C (after Pearson *et al.*, 1978)**

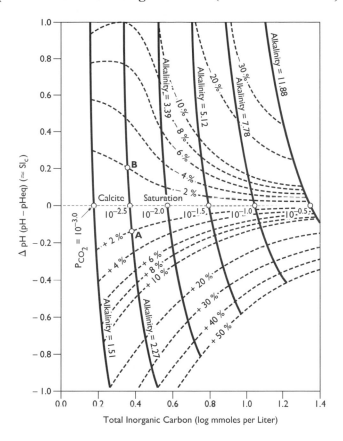

1.2.3   *Effect of changing redox conditions*

The effects of changes in redox conditions mainly concern sample oxidation, since that is the most currently encountered problem. The phenomenon may occur in all the phases of the sample treatment, from core sampling to water extraction, and even before water analysis. The stability of the redox potential depends on the amount and type of potential determining redox couples (Pitsch *et al.*, 1995a). In sedimentary systems, the redox buffering capacity of the rock may be attributed to iron-bearing minerals (Beaucaire *et al.*, 1998). As soon as water is isolated, the solution might be poorly buffered and, consequently, very sensitive to environmental oxygen.

We have already examined the effects related to the bad conditioning of samples during storage (see Chapter II § 2). Most of the water-extraction methods are now used in a nitrogen atmosphere. Some squeezing-cell chambers are completely built in a nitrogen chamber. Simple centrifugation is also subject to increase oxidation, while the centrifugation with heavy immiscible liquids protects the sample far better and is most suitable for trace-element analysis (Batley and Giles, 1979). Leaching first showed which were the serious consequences of oxidation, namely a sulphate increase if pyrite is present, and a lowering of the pH, thus affecting the carbonate system (Oscarson and Dixon, 1989).

Two major consequences are related to oxidation. One concerns the solid phases that are oxidised (generally all sulphides are very sensitive), or whose solubility changes. In fact, the sulphate increase tends to lower the pH, thus affecting the carbonate system (solution degassing, calcite dissolution) and the clay minerals (modifying their stability). The second is related to a change in the functional groups on the organic matter and a modification of sorbed ions on clay surfaces, leading to a different behaviour with respect to the ions and metals in solution (Sposito, 1984). The processes are well explained, but their effects are difficult to evaluate quantitatively, due to the lack of thermodynamic data.

Oxidation artefacts are very well documented in high water-content systems, such as marine and lake sediments. Those seem to affect primarily trace elements (phosphates and iron are strongly depleted in the solution) and, to a minor extent, silica (Bray et al., 1973). The decrease in PO_4^{2-} and Fe^{2+} are related to the precipitation of vivianite in iron-rich sediments or to the precipitation of iron hydroxides with subsequent phosphate adsorption, while for silica, it should be related to changes in the adsorption of silica on the clay particles. According to Lyons et al. (1979), the changes in pH, SO_4^{2-}, Ca^{2+}, Mn^{2+} and alkalinity were not significant. On the other hand, de Lange et al. (1992) reported an oxidation artefact for K^+ and possibly B^-, the nature of which is discussed without reaching a definite conclusion.

Lessard and Mitchell (1985) observed on quick clays that, even if the water content stayed constant, calcium, magnesium and sulphate contents increased by several folds. The oxidation of iron sulphide would result in the formation of iron hydroxides and sulphuric acid, while the oxidation of the organic matter would produce carbonic acid. That would lead to the dissolution of calcium carbonate. One final effect would be the increase in sodium and potassium in porewater through cation-exchange processes, in response to the input of divalent cations.

In indurated clays, due to the very low permeability, oxidation artefacts seem to proceed much slower. Nevertheless, the water/rock ratio being lower, their effects on the water composition are very important. Due to the number of reactions that might be involved, it is virtually impossible to correct the water compositions obtained in oxidised conditions (Pearson et al., in preparation), and data are normally discarded. That is why special attention should be paid to avoid that artefact, if reliable results are to be expected.

1.2.4   Effect of changing temperature conditions

The influence of changes in temperature on the solution composition were first demonstrated in the framework of deep-sea drillings, where samples where subject to ~ 20°C warming from their initial conditions before being squeezed. That temperature effect seemed related to the type of sediment involved, clays and marly sediments being the more sensitive (Sayles et al., 1976).

Temperature was shown to increase potassium concentrations in the extracted solutions at a rate of 1% per 1.5°C (Mangelsdorf et al., 1969). Bischoff et al. (1970) also found a decrease of magnesium and calcium (1.4% and 2.8% respectively each 10°C), and an increase of chlorine (0.8% per 10°C) with temperature, suggesting in addition a change in the bicarbonate content to balance the electroneutrality of the solution. The authors explained the phenomenon by a modification of the cation/anion exchange capacity of the clay in relation with temperature. Silica (Fanning and Pilson, 1971) was also found to increase rapidly (2.6% per degree), while pH seemed to increase much less in the same conditions and no change was observed for interstitial phosphate and salinity. In that case, a change in mineral solubilities or a release of adsorbed silica had to be evoked to explain the observation, as also suggested by the reversibility of the reaction.

According to Kriukov and Manheim (1982), most of the observed changes in composition are reversible and may be corrected by storage for a couple of hours at *in-situ* temperature prior to water extraction. Nevertheless, if calcite precipitation occurs, the temperature-re-equilibration may not be sufficient to restore initial conditions (de Lange, 1992).

De Lange *et al.* (1992) compiled a comprehensive review of the possible artefacts related to temperature and storage conditions of marine sediments. In that study, the authors interpolated the data obtained on piston cores based on concentration differences of samples squeezed at 4 and 22°C. That was based on the assumption that the boron content of sea porewater is conservative and its deviation is linear with the encountered temperature increase. If the temperature dependency of cations were due to temperature-dependent cation-exchange equilibria between the sediment and the porewater, then, for each cation, it should be possible to correct the measured composition taking into account the calculated temperature deviations. After an extensive discussion, the authors concluded that that correction may not be readily made for sediments of highly-variable composition.

An attempt to determine the temperature-induced artefacts has been recently conducted to correct boron content and isotopic data (You *et al.*, 1996). The study showed that the partition coefficient of boron is a function of temperature, pH and sediment mineralogy. Using a barium-10-spiked synthetic solution with major-element composition similar to that of *in-situ* porewater, samples of sediment mainly composed of clay minerals were processed in a rocking autoclave over a temperature range from 25° to 350°C at 800 bars. *Kd* values were determined for boron and barium-10/barium-11 distribution in the extracted solutions. That allowed to derive empirical equations relating *Kd* to temperature and pH for different mineralogies, while the $\delta^{11}B$ of porewaters was corrected assuming the fractionation factors at different temperatures. Although that study concerned particular conditions of high temperature and pressure, it highlighted a possible method to correct temperature induced artefacts.

The issue of changing temperatures for low water-content systems has not been considered yet, although the differences in temperature are not as serious as in the case of marine sediments. Nevertheless, most of the squeezing cells and centrifuges are currently equipped with temperature-control devices ensuring that the extraction procedure is conducted at least at a constant temperature.

Core freezing to preserve water-sample compositions has been proven useful (Griffault *et al.*, 1996), provided that special attention is paid to control the cold chain and avoid ice sublimation by protecting the sample with appropriate coating. It should be remembered that biological activity is strongly increased by defrosting, and that may lead very quickly to the modification of the original content and type of organic matter.

1.2.5 *Ion exchanges*

Ion-exchange processes are launched whenever the solution composition in the clay/water system is modified. That may happen following the introduction or loss of gases (oxidation, degassing of CO_2), a change in temperature, a change in the water/rock ratio, the dissolution or precipitation of soluble phases, the decomposition of the organic matter, etc. It is obvious that the phenomenon has to be considered in each extraction procedure, and may not be excluded *a priori*. In order to evaluate and possibly correct ion exchanges, either the "true" interstitial-solution composition is needed (*e.g.*, by comparing it with piezometer flowing water), or the ion occupancies on the adsorption sites and the way those sites react to the induced disturbance need to be known.

A first approach to the problem consists in evaluating the clay surface charge and the ion-exchange capacity (Sposito, 1984; Hochella and White, 1990). As we have seen in Chapter I, the existence of negative charges on the clay surfaces leading to cation adsorption is due to isomorphic substitutions within the structure, broken bonds at the edges and external surfaces, and dissociation of hydroxyl groups. Anion retention may occur due to the replacement of hydroxyl groups and adsorption on the edges of the minerals. In general, ion-exchange processes are pH– and ionic-strength dependent.

The CEC and the anion-exchange capacity (AEC) measurements usually involve saturation with an index cation or anion and determination of the amount adsorbed. The measurement is performed at a given pH of 7, and results are expressed in milliequivalents (meq) per 100 g of rock (Rhoades, 1982; Wilson, 1987). A number of cations have been used (ammonium, calcium, barium, silver-thiourea, cobalt-hexamine, ethylenediammine, and others) leading sometimes to very different results. In fact, the negatively-charged surface does not display the same affinity for all cations, but exhibits selectivity depending on the ionic charge, size and state of hydration. In general, it is observed that cation-exchange selectivity increases with cation charge, and, for a given charge, with ionic radius. As a consequence, the order of increasing affinity, according to Swartzen-Allen and Matijevic (1974), is:

- – for alkali-metal ions: lithium < sodium < potassium < rubidium < caesium,

- – for alkaline earth ions: magnesium < calcium < strontium < barium.

AEC measurements are even more subject to experimental artefacts, as anions normally show a negative adsorption (de Haan, 1965). Measurements indicate that, depending on the type of clay and on the pH of the solution, the selectivity sequence for anions is:

nitrate < chloride < fluoride = sulphate < molybdenate < selenite < arsenate < phosphate

The measurement is complicated by the fact that aluminium and iron may form insoluble salts with the exchanging anions, making it uncertain whether the depletion in the solution is due to adsorption or precipitation. Solutions of barium chloride (Gillman, 1979) or ammonium chloride (Wada and Okamura, 1977) allow the simultaneous measurement of CEC and AEC. For the large uncertainties in the results obtained with different solutions (Van Olphen and Fripiat, 1979), the type of measuring method used to evaluate ion-exchange capacity should always be specified together with the results.

Those characteristic parameters proceed from a very simple representation of the clay surface as an ion exchanger of a single type. The variability of the experimental values for the same material may be explained by the fact that various types of exchanging sites actually exist on the clay surface (see Chapter I § 2.3.1). Even if in restricted application fields a simplified model may provide interesting information (Pitsch et al., 1992; Bradbury and Bayaens, 1998), recent work on pure clayey phases showed that the multi-site character of the clay surface may be evidenced by the thermodynamic modelling of sorption isotherms (Gorgeon, 1994; Gaucher et al., 1998). Using that approach, cation-exchange capacities and selectivity may be determined for each site.

1.2.6 *Effect of salt dissolution and precipitation*

Dissolution or precipitation of solid phases may be observed when parameters such as temperature, pressure and water content change in the clay/water system. As a consequence, all the extraction techniques may be affected by those phenomena.

The dissolution of soluble salts has been extensively studied for leaching, and constitutes one of the major arguments against the use of that technique (Manheim, 1974). In fact, it has been proven that, as soon as the interstitial solution is modified by dilution or addition of ions, ions start to exchange with the solid surface, with largely unpredictable consequences on porewater chemistry, unless a detailed sorption study has been previously conducted.

The solid phase may be characterised to some extent by X-ray diffractometry and microanalysis. Those techniques, provided that the sample preparation is not introducing artefacts (Wilson, 1987), may provide reliable information on the presence of soluble salts in the rock matrix. The effects of dissolution may be partially corrected if the thermodynamic properties of the dissolving phases are known and the ion-exchange processes mastered (Bayens and Bradbury, 1991).

Salt precipitation, on the other hand, has been poorly studied because of the difficulty to determine newly-formed phases, often in very small amounts and sometimes poorly crystallised. That phenomenon is potentially very active and may have serious consequences on both the chemistry and the isotopic composition of the extracted solutions. That argument is supported by the strong salinities observed in porewater chemistries ($I = 0.4$ mol kg^{-1} at Mont Terri), that are likely to increase even more when water is extracted. In addition, salt precipitation is likely to occur depending on the local (within the sample) conditions of supersaturation (Fritz and Eady, 1985). Ongoing research at CIEMAT, in the framework of the FEBEX project of the European Community, will try to observe that phenomenon by scanning-electron-microprobe (SEM) examination on squeezed core samples.

Evidence of an isotopic shift most probably related to the precipitation of newly-formed phases was reported in Poutoukis (1991). Isotopic analyses were performed on naturally-dripping water in a potash mine in France, and on porewater extracted by vacuum distillation from marly layers interbedded in the salt formation. The salt content of those solutions was extremely high: they were brines saturated with respect to halite and sylvite. According to the data interpretation, the isotopic shift observed between vacuum-distilled porewaters and corresponding dripping waters may be related to the precipitation of highly hydrated salts (carnallite [$MgKCl_3 \cdot 6H_2O$], bischofite [$MgCl_2 \cdot 6H_2O$]) during distillation, all leading to a depletion in heavy isotopes of the extracted solution (Figure 49). That example might seem far-removed from what is likely to be found in clay rocks, but it showed a phenomenon that may potentially occur at different scales and with more limited effects.

France-Lanord (1997) found, for samples coming from the Meuse investigation site, that strongly-bound water had a deuterium value enriched of approximately 10‰ with respect to free water. On Tournemire samples, Ricard (1993) reported a deuterium enrichment exceeding 60‰ of strongly-bound water (recovered by distillation at 100°C of a rock sample previously distilled at 60°C) with respect to water obtained by distillation at 60°C. Nevertheless, due to the small quantities of the first, the cumulative isotopic composition would not be sensitive to that contribution. In both cases, the small amount of strongly-bound water extracted did not allow the simultaneous measurement of the oxygen isotopic composition. In the case of the Meuse samples, it was also observed that the porewater representative data fall above the meteoric water line. Mineral neoformation was evoked as a possible phenomenon causing the observed isotopic shift: possible newly-formed phases are smectite, silica, carbonates or hydroxides, all affecting the oxygen isotopic composition to a greater extent than hydrogen. France-Lanord (1997) calculated that the recrystallisation of 2% of smectite may induce the observed shift off the meteoric water line.

A similar shift off the meteoric water line was also observed at the Mont Terri site and was, in that case, attributed to an artefact of the vacuum-distillation technique (Pearson et al., in preparation).

Figure 49. **Isotopic composition of brines and porewaters extracted from marls by vacuum distillation (after Poutoukis, 1991)**

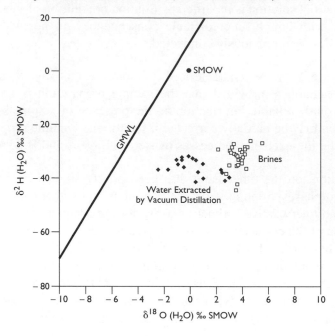

Although the precipitation of minerals during the extraction may be evoked to explain the shift, it is unlikely that clay precipitation be responsible for it. A simple calculation using porewater chemistry showed that the salinity would not be sufficient to precipitate enough clay minerals to account for the observed isotopic shift; it would therefore be necessary to consider other precipitating phases displaying a much higher fractionation factor (Zheng, 1993).

Precipitation of newly-formed phases or recrystallisation and re-equilibration of clay minerals are known to occur during diagenesis. Hower *et al.* (1976) studied the mechanism of burial metamorphism of a type of shale. They found the fine fraction (< 100 nm) of the shale to be constituted by almost pure illite/smectite, whose composition would change with depth from a 20% illite to an 80% illite. That change was also detected in the mineralogical composition, where calcium and sodium, present as exchange ions, would decrease with depth in accordance with the reduced exchange capacity of the smectite when converted to illite. Magnesium, iron and silicon, present in the smectite, would decrease with increasing depth. The authors also suggested that magnesium and iron lost from the smectite fraction would take part in the formation of chlorite, as burial depths increase.

In summary, a precipitation or recrystallisation of clay minerals, most likely occurring during diagenetic processes may be responsible for the observed isotopic shift above the meteoric water line in some indurated clays. In our opinion, that topic requires more investigation, since it may be an indication of a naturally-occurring phenomenon that prevents the interpretation of the porewater isotopic compositions in terms of origin and age.

1.2.7 *Incomplete water extraction*

Incomplete water extraction is a phenomenon mainly affecting the isotopic composition of the solutions. We have already discussed that effect when addressing the vacuum and azeotropic distillation techniques. We have seen that the observed effects may be sometimes corrected, provided that a satisfactory model describing the fractionation during extraction is developed. In fact, it is observed in most cases that incomplete extraction does not follow a simple Rayleigh distillation model (see Chapter II § 3.4 for more detailed discussion).

Also associated with that topic, but much less investigated, is the possibility that incomplete water extraction may be related to the salinity of the interstitial solution, via two main mechanisms: precipitation of newly-formed (hydrated) phases, and water retention in the hydration sphere of the cations. The first topic has already been discussed.

We have seen in Chapter I that some cations form very stable hydration complexes. There is considerable evidence of the reduced mobility of the water molecules in saline solutions. From the thermodynamic point of view, brines display a water activity that differs from unity[12]. The relevance of that phenomenon has been studied especially from the isotopic point of view (Sofer and Gat, 1972; 1975; Yechieli et al., 1993). The conventional carbon-dioxide-equilibration procedure for oxygen-18 analysis in aqueous solutions yields $\delta^{18}O$ values that depend on the water activity of the solution. For dilute solutions, activity and concentration are very close and the measured delta values do not need to be corrected (Jusserand, 1979). For brines, it is necessary to introduce a correction related to the salting-out effect of different cations (Sofer and Gat, 1972):

$$\frac{\delta_O - \delta_m}{\delta_m + 1000} * 10^3 = 1.11\,M_{Mg} + 0.47\,M_{Ca} - 0.16 M_K$$

where δ_O is the value on the concentration scale, δ_m is the measured value, and M is the molality of the ion. From that expression, it is clearly seen that the magnesium concentration plays a significant role in the deviation of water from the ideal behaviour. That is because magnesium forms a very stable hydration complex with water molecules (Taube, 1954; Ohtaki and Radnai, 1994).

A simple calculation for the most saline porewaters (Opalinus clay) shows that the correction for oxygen-18 would be within the analytical error for the analysis. Nevertheless, it is suggested that the salinity found in interstitial solution may be responsible for the difficulty to strip water molecules surrounding cations. In addition, it may also be responsible for some fractionation affecting the different cations during solute extraction, their mobility being more related to their hydration radii that to the dimension of the cation itself. That subject has been poorly studied, both from the experimental and the theoretical point of view, and in our opinion, should be addressed in the near future.

1.3   Current understanding

After having discussed all the extraction methods, their known artefacts and the physical principles that may lead to chemical and isotopic fractionations during the water-sample collection, there is little doubt that many arguments may be raised against all the investigations conducted with those methods and the results obtained.

---

12.   The activity a_i characterises the chemical reactivity of the i species. For a solute i, the relationship with the measured concentration is $a_i = c_i\gamma_i$, where c_i is the concentration in $mol\cdot l^{-1}$ and γ_i is the ionic activity coefficient. That coefficient may be related to the ionic strength of the solution. For the solvent (i.e., the water), the activity is the relationship between the actual partial water pressure over the solution and the partial water pressure over pure water at the same temperature. Its activity in a solution is therefore always lower than 1.

Naturally, the basic problem relates to the presence of different types of water in the clay/water system. From the hydrogeological point of view, only the free water (in amount and composition) is relevant, because it represents the fraction possibly mobilised under given hydraulic conditions. Nevertheless, in the type of rocks considered in this review and featuring very high clay content and very low water content, the amount of free water is probably so very small that each attempt to extract it will inevitably have to deal with the strongly bound water as well. That may be due to the dishomogeneity of the water distribution inside the sample, affecting the local conditions of water availability, as well as to the slow movement, due to the diffusion of the water molecules themselves. Attempts to evaluate quantitatively the amounts of free versus bound water, also using adsorption-desorption isotherms, are not routinely conducted.

Ideally, the approaches that have been adopted so far either claim to extract only the relevant type of water or try to extract all the water and calculate the composition of porewater, assuming the behaviour of the water/rock system. Unfortunately, both those options are not verified, and in each study case we have evidence of the chapterial extraction of different types of solutions, the relative amounts of which remain unknown.

What information may we consider as reliable then? In our opinion, very few. Those are mainly distribution profiles across the studied clay formations for chloride obtained by leaching, deuterium obtained by distillation (provided that no serious salinity differences are detected in the porewaters) and, to some extent, noble gas measurements. We will try to justify our statements.

Chloride has always been considered the only ion that may be obtained in correct concentrations by leaching, provided that the absence of soluble salts containing chloride is proven. That is the common case, as chloride is rarely included in the mineral crystal lattice, except in evaporites originating at high brine concentrations. That is why that ion, coupled with bromide, is used to distinguish between primary brines (deriving from evaporated seawater) from secondary brines (derived from evaporite leaching). Chloride is also not extensively adsorbed by clay, as most of the negative ions, especially at basic pH (de Haan, 1965). The leaching should be conducted carefully, avoiding environmental contamination or loss of fairly volatile chloride, especially if isotopic studies are planned. Chloride is released quite slowly in the leaching solution, and that process should be checked by testing the electric conductivity of the solution, or ensuring better that no more chloride is released. According to the granulometry of the sample, chloride extraction should take only a few days.

Some cases have been reported where chloride leached from the rock seemed to be dependent on the selected solid/liquid ratio (Figure 50). That was observed especially in environments displaying a low chloride content in porewater (Moreau-Le Golvan, 1997; Cave et al., 1997). The reason for that behaviour is not quite well understood, since no fluid inclusions were observed in the rock. Chloride analyses of the whole rock (15 times greater than porewater values) suggested that chloride may be retained at the clay surface by adsorption and only released with difficulty. If that is verified for the case study, the same solid/liquid ratio should be used on all samples in order to draw the distribution profile across the formation. In that case, chloride data have to be interpreted in relative rather than in absolute terms.

Recent studies on chloride distribution in indurated clay formations show that the content may often be explained by using a diffusion model (Gard, Meuse, Mont Terri). Coupling of the distribution data with chlorine-36 and chlorine-37 isotopic data may provide additional information. Chlorine-36 may be used to evaluate the chloride residence time (Moreau-Le Golvan, 1997) and to check its "secular" equilibrium with the *in-situ* neutron flux calculated for the formation rocks (Andrews et al.,

Figure 50. Chloride contents back-calculated to porewater content as a function of the solid/liquid ratio (after Moreau-Le Golvan, 1997)

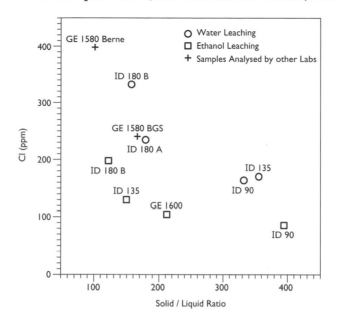

1989). The distribution of chlorine stable isotopes is increasingly studied because of the information that may be deduced about the origins of the chloride source and processes affecting the movement of chloride ions. Diffusion seems to be the predominant mechanism, but chloride is known to fractionate during ultrafiltration (Kaufman *et al.*, 1984; Phillips and Bentley, 1987). The "age" of chlorine in the formation may be calculated, assuming the porosity and tortuosity parameters of the matrix, by computer codes. Those data may be compared with the calculations performed for deuterium-distribution profiles, remembering that, according to some experimental data (Gvirtzman and Gorelick, 1991) and the Donnan principle, chloride is possibly moving inside the formation much faster than water molecules.

Concerning isotopic data, only deuterium-distribution profiles may be considered as quite reliable, provided that no major changes in the porewater salinities across the formation are observed, that the same preparation and distillation procedure has been used for all samples and that the extraction yield is carefully checked. It has been proven, in fact, that deuterium is much less affected by water-extraction artefacts than oxygen-18 (Moreau-Le Golvan, 1997). In addition, the possible modifications due to incomplete water extraction, either related to the precipitation of newly-formed phases, or to the hydration sphere of cations are less influent on the deuterium values, and usually fall within the analytical error. Even if those phenomena cannot be excluded *a priori*, they are likely to affect all the samples the same way, if salinities do not differ substantially. Results may only be interpreted in relative terms. In other words, the obtained values should not be considered as the "true" porewater values, but be compared to each other. Distribution profiles across the formation may be drawn (Cherry *et al.*, 1979; Desaulnier *et al.*, 1981; Falck *et al.*, 1990; Moreau-Le Golvan, 1997), and computer codes allow to calculate the time required to establish that profile (based on the porosity and tortuosity data of the clay rock) and the predominant water-movement mechanism (Neuzil, 1995; Darling *et al.*, 1995).

Finally, noble-gas measurement and distribution profiles are potentially very informative. Fewer artefacts may affect those measurements, provided that special attention is paid to the careful conditioning of the sample soon after drilling. Noble-gas content and isotopic distribution allow once more to provide an age estimate (Osenbrück *et al.*, 1998).

We are conscious of that rather restrictive interpretation. On the other hand, we have extensively discussed all the possible artefacts, and until a deeper comprehension of the clay/water system exists or that new techniques eliminate artefacts, the amount of actual data that may be readily collected and interpreted is very limited. Nevertheless, all the studies that have been conducted up to now on clay-rich media should not be discarded, since useful information on porewater chemistries may still be obtained from them through extensive modelling based on the thermodynamic properties of the clay/water system.

2. GEOCHEMICAL MODELLING

2.1 Dissolution-precipitation

Since the compilation of computer codes with thermodynamic databases for aqueous speciation and mineral solubilities, geochemical modelling has been recognised as a very powerful tool in hydrogeochemical investigations. Three main objectives may be pursued, that are, in increasing order of complexity, the evaluation of the chemical data, the identification of factors controlling the water composition, and the behaviour prediction of the water/rock system.

The first check on the quality of chemical data is normally performed using the electroneutrality equation, stating that:

$$\Sigma[A^-] = \Sigma[C^+]$$

where $[A^-]$ and $[C^+]$ are the anions and cations concentrations expressed on the equivalent scale. It is generally accepted that a complete chemical analysis ranges in the ± 5% error, but "extreme" solutions (very low or very high in TDS, very high in organic-matter content, etc.) may fall outside the range. That is mainly because analytical uncertainties are greater and/or because other unaccounted for charged species are present (*e.g.*, organic acids). The electroneutrality equation is used by most computer codes for compensating analytical data.

Besides analytical difficulties, data errors may also be due to artefacts related to sampling and water-extraction stages. That aspect has already been discussed, when appropriate, for each case study and each technique used.

Dissolved elements in natural waters may be divided into two categories:

– "free" elements, like chlorine, whose concentration is not limited by any chemical equilibrium, because they do not react with insoluble organic matter and may only be incorporated in very soluble salts;

– "controlled" elements, whose solubility is limited by the precipitation of a poorly-soluble mineral: for example, the calcium concentration is often limited by the solubility of calcite.

The identification of the controlling solid phases is achieved through the calculation of the saturation indexes. Schematically, for each dissolving salt or mineral, an equilibrium constant (solubility product) is defined for each temperature:

$$A_n B_m \Leftrightarrow nA^- + mB^+$$

$$K_s(T) = (A^-)^n \cdot (B^+)^m$$

where the activities of the different species are shown in parentheses. When a mineral or a salt is in contact with water, it will dissolve until reaching its solubility product. During that process, its ion-activity product (IAP) is defined as:

$$IAP = (A^-)^n \cdot (B^+)^m$$

As long as *IAP* is lower than K_s, the solution is undersaturated with respect to the solid phase (*i.e.,* the solid will keep dissolving until *IAP* is equal to K_s). When saturation is reached, the activities in the solution will stay constant. A third case may be observed where *IAP* is higher than K_s, corresponding to a metastable condition where precipitation should occur in order to lower *IAP*.

The saturation index (S.I.):

$$S.I. = \log \frac{IAP}{K_s}$$

indicates the degree of saturation of the solution with respect to the considered solid phase, 0 corresponding to the saturation point. In other words, unity saturation indexes point at the solid phases controlling the water composition. Besides, oversaturation may not indicate a metastable *in-situ* condition, but may be due to analytical and/or computation artefacts, possibly related to an inconsistency in the data set. For example, degassing of carbon dioxide normally leads to an oversaturation in carbonate phases (Pearson *et al.,* 1978), due to the retardation of carbonate precipitation. Once again, when that type of artefact was known to be related to the extraction technique applied, it was described in the relevant section of Chapter II of this report and will not be discussed again. However, if no artefacts may be detected, an oversaturation may indicate that the assumption of equilibrium existing in the system is not valid. In fact, in natural systems, precipitation may be inhibited by a number of factors, including kinetics and the presence of complexing organic matter that is often not taken into account in data computing.

Once the data set and the calculation agree, modelling allows for different prediction levels. Those range from the evaluation of unmeasured properties of the system (*e.g.,* oxido-reduction potential), to the evaluation of the effects of an induced disturbance (*e.g.,* heating, freshwater or contaminant intrusion). That ultimately points in the direction of the safety assessment, allowing the prediction of the radionuclide behaviour in the considered water/rock system.

2.2   Sorption

Models have been established to analyse and quantify the adsorption at the clay/solution interface, according to the different parameters of the system considered. Empirical adsorption models, based on mathematical fitting of sorption isotherms, as well as thermodynamic models, such as the surface complexation or ion-exchange models, are described in the literature. It is not the purpose of this report to review all of them, and we will only briefly describe some basic principles. A good review may be found in Sposito (1984) with an accurate mathematical treatment of the different models.

Empirical models are based on the concept of surface excess (q) (*i.e.,* the number of moles of the substance sorbed per unit mass of clay). Batch experiments are carried out with the solid immersed in an aqueous solution containing a number of moles of the sorbing substance at controlled temperature and pressure. The resulting adsorption data may be plotted against each other, with q as a dependent variable of c, the concentration in the solution, in graphs known as "adsorption isotherms". The graphs may be classified according to their initial slope (S–, L–, H–, C–curves) reflecting the substance affinity for the solid phase at various concentrations in the contacting solution (Travis and Etnier, 1981). The most common isotherm in soil chemistry is the L–curve (Figure 51), characterised by an initial slope that does not increase with the concentration of the substance in the solution. That property is the result of a high relative affinity of the solid phases for the substance in solution, coupled with a decreasing amount of adsorbing surface since it is saturated. Mathematically, that curve can be described by the Langmuir equation:

$$q = \frac{bKc}{1 + Kc}$$

where b and K are adjustable parameters, b representing the value of q, that is asymptomatically approached when c becomes large. K determines the magnitude of the initial slope of the isotherm. Those two parameters may be obtained accurately by plotting the distribution coefficient K_d:

$$K_d = \frac{q}{c} = bK - Kq$$

against the surface excess q in a linear graph with a slope of $-K$ and an x-axis intercept equal to b (Figure 51).

It is not uncommon to observe that the graph is not a straight line, but a curve with a negative concavity. That can be described as fitting with a two-term series of Langmuir equations:

$$q = \frac{b_1 K_1 c}{1 + K_1 c} + \frac{b_2 K_2 c}{1 + K_2 c}$$

where b_1, b_2, K_1 and K_2 are adjustable parameters (Veith and Sposito, 1977).

Alternatively, the van Bemmelen-Freundlich isotherm equation may be used, as follows:

$$q = Ac^\beta$$

where A and β are positive, adjustable parameters and β is forced to lie between 0 and 1 (Sposito, 1980).

Surface-complexation models and ion-exchange models are based on thermodynamics. The firsts are also called "microscopic models" or "molecular models", because they make hypotheses on the charge distribution at the clay/water interface, concerning the chemical reactivity of the solid surface. On the other hand, ion-exchange models are also defined as "macroscopic models", because they describe the distribution of the elements considering that the counterion/clay system is electrically neutral. It can therefore be described as a whole, without specifically accounting for interfacial processes. Both types of models consider a limited number of reactive sites on the clay surfaces, that are responsible for ion retention. We will not derive the complete mathematical expressions for those models, but we will briefly describe some of them.

In surface-complexation models, the clay surface is modelled as a plane, with electric charges distributed on its surface according to Gouy-Chapman and Stern-Grahame's double-layer theory. The electric layer may be subdivided in a compact layer associated with an external diffuse layer. In order to balance the solid-surface charge, counterions accumulate in the diffuse double layer. Only ions showing a very strong affinity for the surface may be specifically adsorbed, thus loosing part of their

Figure 51. General classes of adsorption isotherms and calculation of the *b* and *−K* parameters according to the Langmuir equation (modified after Sposito, 1984)

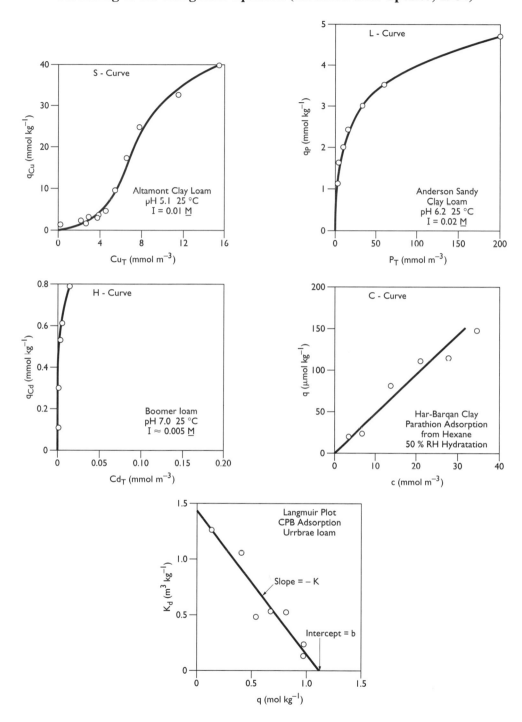

hydration shell. The other counterions are located far from the surface, either in the plane separating the compact from the diffuse layer, also called the "outer Helmholtz plane", or in the diffuse layer. Three models are the most currently used and depicted in Figure 52, together with the potential variation as a function of the distance from the solid.

Figure 52. **Surface complexation models: (a) = diffuse layer; (b) = constant capacitance; (c) = triple layer (after Gorgeon, 1994)**

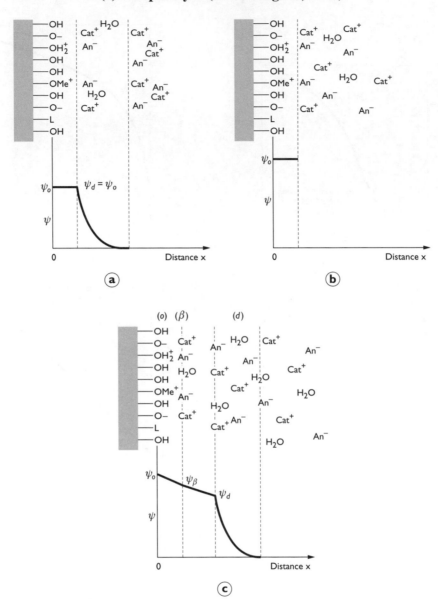

The diffuse-layer model (Stumm *et al.*, 1970) considers that specifically sorbed ions are located on the solid surface (plane 0) and contribute to the surface charge, σ_0. That charge is compensated by the counterions in the diffuse layer, and the compact layer does not include ions. The potential decay in the diffuse layer follows an exponential curve. The existing relationship between σ_0 and the surface potential Ψ_0 takes into account the ionic strength of the solution.

The constant-capacitance model (Schlinder and Kamber, 1968; Schlinder and Gamsjager, 1972) considers that the double-layer thickness may be described as a plane, corresponding to the outer Helmholtz plane. Consequently, the charge density at the surface is proportional to the potential $\sigma_0 = C\Psi_0$, where C is the condenser capacity. That relationship is only valid for solutions with high ionic strenghts or solids with low surface potential (< 25 mV).

142

Both those models only consider inner-sphere complexes. In addition, the influence of the ionic strength is not very well accounted for. The triple-layer model (Hayes and Leckie, 1987) considers another plane (β), also called the "inner Helmoltz plane", where ions showing lower affinity with the solid and forming outer-sphere complexes are located. That model also describes the stronger influence of ionic strength on outer-sphere than on inner-sphere complexes, as shown experimentally.

Ion-exchange models have been established to describe the behaviour of synthetic ion-exchange resins and further extended to natural inorganic exchangers, like zeolites and clays. According to that theory, a solid phase immersed into a liquid phase develops a positive or negative electric charge that is exactly compensated by adsorbed ions. The solid/counterion pair is considered as the ion exchanger and shows no net electric charge. Under those premises, modelling does not require any knowledge of the actual charge distribution at the solid/liquid interface, nor needs extra-thermodynamic hypothesis. The model considers those surface regions, where electric charges are located, as the ion exchange sites. Since different types of exchange sites may be identified on a clay surface, as described in Chapter I of this work, the correct modelling of ion-exchange properties of a clayey material therefore implies to consider them as multi-site exchangers. For example, a cation exchange reaction on the Xi sites is written as follows:

$$m\{(Xi^-)_n\, N^{n+}\} + nM^{m+} = n\{(Xi^-)_m\, M^{m+}\} + mN^{n+}$$

where Xi^- is a site with one negative charge, while N^{n+} and M^{m+} are the exchanging ions and the corresponding equilibrium constant, respectively:

$$K^{io}_{Nn+/Mm+} = \frac{\left(\left(Xi^-\right)_m M^{m+}\right)^n \left(N^{n+}\right)^m}{\left(\left(Xi^-\right)_n N^{n+}\right)^m \left(M^{m+}\right)^n}$$

If activities are replaced by concentrations in both phases, the following selectivity coefficient between N^{n+} and M^{m+} is obtained:

$$K^{i}_{Nn+/Mm+} = \frac{\left[\left(Xi^-\right)_m M^{m+}\right]^n \left[N^{n+}\right]^m}{\left[\left(Xi^-\right)_n N^{n+}\right]^m \left[M^{m+}\right]^n}$$

Each site on the clay is characterised by an individual cation-exchange capacity (CECi), and the total capacity of the material is the sum of those individual capacities:

$$CEC = \Sigma\, CECi$$

Those capacities are intrinsic parameters of each specific mineral.

The model appears suitable to describe metal-ion adsorption. It reduces the number of assumptions needed for modelling because of the electroneutrality assumption for both phases (Gorgeon, 1994), but it does not obviously account for the electrokinetic properties of the clay particles.

2.3 Requirements for geochemical modelling

In order to perform geochemical modelling, a conceptual model is needed. Computer codes are mathematical tools based on pre-existing concepts of chemistry solutions and/or water/rock-interaction schemes. Each model and its corresponding computing code require different parameters, depending on their objectives (McEwen *et al.*, 1990). Basically, water analysis and mineralogical composition of the host rock are concerned.

For a reliable aqueous speciation, water analysis should be as complete as possible and include pH and Eh measurements. Water composition should be checked for possible analytical errors. That is why repetitive analyses are useful for detecting the influence of time, space or the sampling and analytical method used. Internal consistency is also checked by redundancy. For example, the carbonate system is defined by different interrelated parameters, (pH, total inorganic carbon, and carbonate alkalinity) that may be analysed separately and cross-checked. Eh as well may be computed starting from different redox couples.

For predictive modelling, information on the water environment is required. That includes the mineral phases that are present and/or reactive with water. Data include the chemical composition of mineral phases or solid solutions, corresponding solubility products, sorption parameters like exchange constants and site capacities. Isotopic compositions, fractionation factors and amounts may also be needed. In order to allow modelling, a number of codes require the solid phases and the dissolved constituents to be the same in number. Information is also required on those ions whose abundance is normally not controlled by a solid phase (*i.e.*, Cl^-, Br^-). Those help in making consistency tests of the measurements and in identifying problems related to the water extraction technique used. In addition, they serve for defining the geochemical porosity and eventually for reconstructing the geological history of the system (*e.g.* chloride/bromide ratios).

If the measurements and the calculations agree, then the sampling and analytical methods together with the conceptual model are verified as an integrated tool and the parameters may be used for predictive purposes. Validation of the model and the methods requires at least verification on various water/rock systems.

2.4 Computer codes

Plummer (1992) has recently reviewed computer codes for geochemical modelling. Besides simple solution-speciation software like MINEQL (Westall *et al.,* 1976), there are two types of approaches for modelling water/rock interactions.

Forward modelling predicts water compositions and mass transfers resulting from hypothesised reactions. It generally begins with a defined water composition and simulates the evolution of water and rock for a set of specified reactions. Recognised codes of that type are PHREEQE (Parkhurst *et al.*, 1980), PHREEQC (Parkhurst, 1995) and EQ3/6 (Wolery *et al.*, 1990). All codes include the possibility to calculate pH and Eh in response to chemical reactions. That is performed by using, in addition to the mass balance and the mass action equations of the speciation calculations, additional assumptions, such as charge balance, conservation of electrons or constant mass of water. If appropriate thermodynamic data are available for the considered environment, forward modelling is well suited for predicting water compositions, mineral solubilities and mass transfers in response to reversible and irreversible reactions, reaction paths, as well as changes in composition resulting from pressure and temperature changes.

Inverse modelling defines the mass transfer from observed chemical and isotopic data. It produces mineral mass transfers that account for the compositional variations observed in the aqueous system. The equations of mass-balance modelling include element mass-balance, electron conservation and possibly isotope mass-balance. The data needed are analytical data for two evolutionary waters and a set of mineral or gaseous phases that are assumed to include all reactants and products in the system. During calculations, the concentration changes in the elements in the evolutionary waters are used to calculate the masses of plausible minerals and gases that enter or leave the aqueous phase to account for the observed changes in composition. Classical examples of such codes are NETPATH (Plummer *et al.*, 1992) and PHREEQC (Parkhurst, 1995).

The use of forward or inverse-modelling codes depends normally on the purpose of the study and on the type of available data. Forward modelling may be used even if there are no aqueous or solid data for the system. It should be selected, for example, when data are available on mineralogy, chemical and isotopic composition of the solid phases, and mineral parageneses are known, but little information is available on the fluid in contact with the mineral assemblage. Alternatively, in the study of deep aquifers, little or no information may be available on mineral transfers, gases and organic matter. In that case, inverse modelling is more appropriate.

In principle, all computer codes should reach the same results, providing that the input data and the thermodynamic database are the same. Inconsistencies in the databases are often evoked as a possible source of discrepancies in the computed models. That derives from the fact that mineral-stability data are usually obtained from various sources and may be inconsistent with the aqueous model. That is why NAGRA, for example, has made a big effort in the implementation of an internally-consistent thermochemical database to be used for all modelling performed in the framework of its investigations (Pearson and Berner, 1991; Pearson *et al.*, 1992).

Even so, unfortunately, discrepancies may be observed in the results obtained on the same data set modelled by different computer codes. That may be due to the fact that both forward and inverse modelling require considerable insight concerning the problem of selecting appropriate reactant and product phases. As an example, Pearson *et al.* (in preparation) report, for the calculation of the porewater composition for the Mont Terri project, that the unique modelled composition of low pH and high pCO_2 was a modelling artefact due to the use of a given computer code. The re-examination of the theory underlying the calculation procedure showed that the procedure was only able to determine a range of water compositions.

3.   INDIRECT APPROACH TO POREWATER COMPOSITION USING GEOCHEMICAL MODELLING

In 1990, Bradbury *et al.* suggested an experimental procedure to investigate the water chemistry, sorption and transport properties of marls. Their idea was to design a set of experiments that may help to derive indirectly the composition of the pore fluid. A synthetic solution, called the LSM solution, would be created to result in "chemical equilibrium" with the rock. The term "chemical equilibrium" indicates, in that case, that the water/rock system would have reached a steady-state condition. Besides, at the timescale of the experiment, neither the state of the rock nor the water chemistry should be a function of time and of the quantity of rock and volume of the solution used for the experiment (Baeyens and Bradbury, 1991).

Ideally, the LSM water should have the same chemistry as the formation water. However, in rocks with very low water content, it is difficult to understand the real meaning of "*in-situ* composition", when "free" water is present in far lesser amount than "bound" water. That methodological approach is completely different from what we have described so far since it encompasses both the technical problem of water extraction and the definition of the amount and characteristics of "inner" and "outer" solutions. Since its design, the approach has been applied on samples of the Palfris marl (Baeyens and Bradbury, 1991; 1994), and is currently used in the framework of the investigations in the Mont Terri project (Bradbury and Baeyens, 1997; Bradbury *et al.,* 1997; Bradbury and Baeyens, 1998; Pearson *et al.*, in preparation).

Basically, Bradbury *et al.* (1990) started from the observation that, although marls were rich in calcite, the aqueous leachate was of the sodium/bicarbonate type. That might be due either to the presence of a soluble sodium-carbonate salt, or to an exchange process occurring with the clay, taking up calcium and releasing sodium. Consequently, they designed a set of experiments consisting of a series of leaching tests, in order to determine the presence of highly-soluble salts (sodium chloride, sodium bicarbonate, sodium carbonate), the CEC of the clays and the ion occupancies on the exchange sites. Two types of extraction experiments were used. One was aqueous leaching performed in a glove box with controlled atmosphere ($pCO_2 \sim 10^{-2}$, $O_2 < 5$ ppm), using liquid/solid ratios of 1:1 and 1:2. The other utilised a calcite-saturated solution containing nickelethylenediammine (Ni-en), a high-selectivity complex that displaces all exchangeable cations from the clay to the solution. That inhibited the calcium/sodium exchange and allowed simultaneously to measure the cation-exchange capacity. The reason for pre-saturating the solution with calcite was that information should be obtained on the origin of some cations such as sodium, by comparing the pH, inorganic carbon and calcium concentrations in the marl-extraction experiments with their respective values in the leaching solution. Some of the results obtained on the Palfris marl (Baeyens and Bradbury, 1991) were briefly summarised in order to give an example of the modelling approach.

By plotting the calcium concentration *versus* the liquid/solid ratio on a logarithmic scale (Figure 53), for both the aqueous leachates and the Ni-en extracts, the following considerations were made:

–   Calcium entered the Ni-en solution. Since the solution is pre-saturated with calcite, calcium may only have been displaced from the clays by the affinity complex. That allows the calculation of the quantity of exchangeable calcium present in the marl, by calculating the release at different liquid/solid ratios (0.9 meq /100 g).

–   As a comparison, the aqueous leachate, showing much lower values than expected for a solution in equilibrium with calcite, indicated that, in that case, an exchange with clays was operating. Calcium was provided by the dissolution of calcite, but was immediately taken up by the clay, releasing sodium. That in turn caused a further dissolution of calcite, and consequently a further exchange, until the system reached equilibrium.

For the other elements, if a logarithmic scale is adopted for their concentration as well, a plot with a slope of –1 would indicate pure-dilution effects (Figure 54). A greater slope (*i.e.,* less negative) would be indicative of precipitation and/or sorption effects, and a smaller slope, of dissolution and/or desorption processes. As an example, sodium shows a slope of –1, which is interpreted as a displacement from the clay component of the marl by the Ni-en solution, except for a small fraction resulting from the dissolution of sodium chloride. Aqueous leachates show a lower sodium content, indicating that not all the exchangeable sodium is displaced, in that case, from the clay to the distilled water.

Figure 53. **Measured calcium concentrations in Ni(en)$_n$ extracts (full circles) and in aqueous extracts (open circles). The hatched region represents the range of calcium concentration of the Ni(en)$_n$ standard solution (after Baeyens and Bradbury, 1991)**

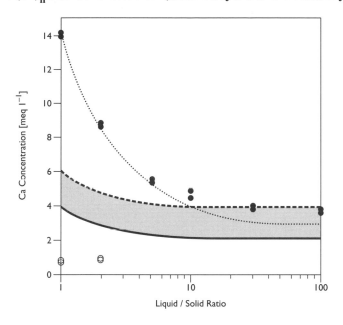

Figure 54. **Measured sodium concentrations in Ni(en)$_n$ extracts (full circles) and in aqueous extracts (open circles) (after Baeyens and Bradbury, 1991)**

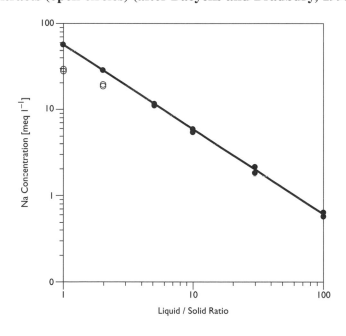

On the basis of the considerations made on all the ions in the aqueous and the Ni-en solution extracts, the quantity of highly-soluble salts (sodium chloride, potassium chloride) is derived, together with the cation-exchange capacity of marl and the ion occupancy on the exchange sites.

One more step is needed in order to obtain an LSM solution that will not change the occupancies on the clay: the calculation of the cation ratios in the liquid phase, that are in equilibrium with known occupancies. That calculation may be achieved using semi-empirical Gapon equations (Baeyens and Bradbury, 1991) or by calculating the selectivity coefficients for sodium and potassium (Baeyens and Bradbury, 1991; Baeyens and Bradbury, 1994). The first approach, based on a very coarse empirical correlation observed in soils (Richards, 1954), leads to rather unsatisfactory results. The second, even if it is based on a quite simplified representation of the clay as a mono-exchanger, is much more satisfactory.

Selectivity-coefficient calculations are based on the reasonable assumption that bivalent ions will exhibit similar affinities for clay and may be treated consequently as being equivalent. The selectivity coefficient of Na^+ with respect to the sum of bivalent cations M^{2+} is defined by the authors, for the given exchange equation, as follows:

$$2Na^+ + \overline{M^{2+}} \leftrightarrow 2\overline{Na^+} + M^{2+}$$

$$^{Na}_M K_c = \frac{N^2_{Na}}{N_M} * \frac{a_M}{a^2_{Na}}$$

where the *bar* denotes the ion adsorbed on the solid phase, a_{Na} and a_M are respectively the activity in the solution of Na^+ and the sum of the activities of the bivalent ions ($a_M = a_{Mg} + a_{Ca} + a_{Sr}$), and N_{Na} and N_M are the equivalent fractional occupancies *i.e.,* the equivalents of sodium (or M) sorbed per unit mass, divided by the cation-exchange capacity, in equivalents per unit mass. As a result:

$$N_{Na} + N_K + N_M = 1$$

The exchange with potassium is treated in the same manner. If sodium and potassium are predominantly present as exchangeable cations on the clay minerals (*i.e.,* if the contribution of soluble salts is negligible), then the sodium and potassium final occupancies at any solid/liquid ratio may be easily calculated from the difference between their initial occupancies and their equilibrium concentrations in the aqueous leachate. Selectivity coefficients with respect to the sum of bivalent cations may then be obtained using the above equations. The calculation of the actual LSM composition is performed using an iterative approach. A complete example of the calculation for the Palfris marl was reported in Baeyens and Bradbury (1994).

The calculations of the LSM composition appear to be relatively sensitive to the selected pH (and pCO_2): an uncertainty of 0.5 pH unit leads generally to 50% changes in ion concentrations. Those parameters should consequently be better controlled, both during *in-situ* and laboratory experiments. Besides, while those calculations are relatively easy and possibly well constrained for major elements, it is difficult to obtain reliable data for minor and trace elements, that, on the other hand, may play a determinant role in radionuclide migration and sorption. For the determination of those elements, the authors suggested to proceed through equilibration tests, readjusting progressively the solution composition. That may also be done for obtaining more reliable pH values (see *e.g.,* the "*in-situ*" pH equilibration test that is currently being performed in the framework of the Mont Terri project).

That modelling approach has been used to obtain porewater compositions in the Opalinus clay (Bradbury and Baeyens, 1998). In that case, the aqueous leachates were conducted with deionised, degassed water at different solid/liquid ratios. The Ni-en extraction solution was no longer saturated with respect to calcite as a result of the difficulty in controlling the calcite precipitation that might occur due to the release of calcium from the exchange complex. Two different cases were examined when calculating the cation occupancies on the clay and selectivity coefficients. In the first case, all the sodium was considered to be associated with the chloride in porewater, and all the calcium with the sulphates. In the second case, sodium was associated with both the dominant anions. The two data sets were not significantly different. Assuming the saturating mineral phases to be calcite, dolomite, chalcedony and fluorite, the calculated fractional cation occupancies and the chloride and sulphate contents (that depend on the assumed porosity value) (Pearson, 1998), one degree of freedom still existed; which implies that a unique porewater composition cannot be calculated. The missing component was pH, or alternatively pCO_2 both of which may be estimated based on an additional reaction. The systems did not seem to be highly affected by the selected value of pCO_2, except for pH and bicarbonate content. The highest degree of uncertainty in the modelled porewater composition was associated with the selected porosity.

The same modelling approach applied to the indurated clays of the Tournemire site allowed the estimation of a porewater composition, that may not be obtained by direct means due to the extremely low water content of the formation (Boisson *et al.*, 1998; De Windt *et al.*, 1998a; Waber and Mazurek, 1998). While in that case, porosity was better defined, the interpretation of the aqueous leachates to determine the readily-soluble salts was not unique. As a consequence, there were some problems in defining the mineral phases controlling the porewater composition, and sulphates were not taken into account in the calculation. In addition, the Tournemire case was very different from the previously-examined Swiss cases in that it has a calcium and magnesium-dominated exchange population, which makes it even more difficult to constrain the system. Nevertheless, the calculated porewater-composition range was in reasonable agreement with the groundwater obtained from a fracture in the Tournemire shale.

In any case, it is worth observing that, even if the simplified clay-surface representation may be accepted to a certain degree for modelling in a very narrow pH range, neglecting H^+ sorption seems to be a much more risky hypothesis, which may explain certain limitations in the predictivity of the model. Indeed, when applying ion-exchange concepts to freshening of aquifers, Appelo (1994) used a simple mono-site model, but had to include proton exchange in order to get representative results.

Another modelling approach that has been proven valid for clay environments was proposed by Beaucaire *et al.* (1995; in press) on the Boom clay, within the work of the ARCHIMEDE programme. Their study relied on regional-groundwater characterisation (Boom-clay porewater and groundwaters from the Rupelian aquifer). The acquisition and regulation mechanisms of the groundwater compositions at the regional scale were considered to be the same, and the fluids to have a common origin. By a careful observation of the correlation between major cations and anions, a mixing between recharge porewater and a marine solution may be identified. However, a simple mixing model between those two end members did not describe the system accurately, thus suggesting that exchange and equilibration with the host rock also occurred. Referring to Giggenbach and Michard's isochemical recrystallisation model successfully applied to the modelling of granitic waters (Grimaud *et al.*, 1990), Beaucaire *et al.* used a dissolution-precipitation model considering a mineral assemblage constituted by kaolinite, chalcedony, calcite, dolomite, albite and microcline in order to predict the water composition. Those phases had all been identified during the mineralogical analysis of the Boom clay. They were not all known to be present in the aquifer matrix, but could be reasonably suspected since the aquifer is located in clayey sands of the same geological group. Those minerals were also selected because their dissolution equilibria are well established. Clay minerals instead may be treated either as defined phases or as solid solutions, thus making it difficult to define thermodynamic constants.

The external variables of the model are temperature and the concentrations of the "free" elements that do not participate in any dissolution-precipitation equilibrium: chlorine as chloride and sulphur as sulphate. At that stage of the modelling, the total inorganic carbon concentration was also considered as non-controlled, because no satisfactory mineral buffer could be identified.

The model seemed to describe accurately the variability of the recognised types of water within the regional scale. Discrepancies between predicted and measured concentrations of major elements were within analytical uncertainties, except for very dilute species; with regard to pH, the deviation was less than 0.3 unit. That good predictivity induced De Windt *et al.* (1998b) to test the model on the Tournemire water collected from a draining fracture of the site and resulted in a quite satisfactory agreement between modelled and analysed water compositions.

Regarding trace elements, that modelling approach proved to be quite impressive in explaining the low oxido-reduction potential in the Rupelian aquifer by a geochemical control through goethite and siderite. In the Boom clay, Eh, iron and sulphur concentrations were determined by an assemblage of pyrite, siderite and goethite: the Eh value calculated by that model only differed by 20 mV from the measured one (Beaucaire *et al.*, 1998).

In the ARCHIMEDE programme, surface-solution equilibrium was also tested on Boom clay and porewater, using the exchange constants defined for pure clay minerals in the natural assemblage (Gorgeon, 1994) and assuming additivity of the exchange properties, including mixed layers. That model allowed to describe the metal-cation fixation on one illite and one smectite site, and the calculation of calcium and magnesium concentrations in the porewater. The computed values were somewhat higher than the measured values, but that modelling approach seems reasonable (Beaucaire *et al.*, in press). Ongoing work on Boom clay is devoted to characterise more accurately its sorption properties in order to refine the model. Besides, the same authors modelled the reaction of the clay/water system to an oxidising disturbance by a combination of dissolution and ion-exchange processes. The latter are easily shown in such a case due to their high velocity.

In fact, both phenomena are known to occur during water/rock interaction, but at different timescales: ion exchanges are quickly established while equilibrium is more slowly attained in dissolution-precipitation reactions. As a consequence, if a groundwater is equilibrated with the host rock, it is necessarily in equilibrium with the rock surface. A second consequence is that disturbances induce immediate surface reactions: for example, water intrusions or leaching experiments may only be correctly interpreted if those processes are taken into account. On the other hand, a pure ion-exchange model of groundwater cannot be totally constrained, since it would involve the exact knowledge of the chemical composition of the rock surface. Conversely, a dissolution-precipitation model only needs to take into account the nature of the dissolving solid phases. Naturally, both aspects need to be considered in the safety-assessment procedure for waste-disposal sites, which means that a considerable work of thermodynamic-data generation has to be done in order to reach reliable conclusions.

4. CONCLUSIONS, RECOMMENDATIONS AND TOPICS FOR FURTHER INVESTIGATION

The problem of extracting solutions from argillaceous formations for geochemical characterisation is a complex one, as expected. For the time being, the presence of different forces arising from the clay/water interaction and influencing the movement of water molecules and solutes prevents the possibility to define experimentally the "true" porewater composition. That composition is needed for several purposes:

- To calculate the corrosion of the canisters and the matrices where radioactive waste will be confined.

- To evaluate the origin, the residence time and the natural processes of water and solute movement through the formation.

- To calculate the speciation and the solubility of phases in order to evaluate the water/rock-interaction phenomena affecting radionuclide migration.

- To foresee the effect of the site water on the engineered barriers of the repository.

Fortunately, not all those purposes require the same knowledge of the porewater composition.

For corrosion studies, the main parameters needed are the total salinity, the oxidising or reducing properties of the solution and the speciation of particular elements (*e.g.,* sulphur). In the host rock, it may be reasonably assumed that, given the high solid/water ratio, the solution composition would soon be controlled by the minerals in the matrix, the composition and reactivity of which are fairly well known.

For tracing back the origin, age and movement of the water, a few techniques have proven reliable. Noble-gas measurements may provide information on groundwater age. Deuterium and chloride, provided that they are interpreted in relative rather than in absolute terms and that porosity characteristics are well known, allow an estimation of the time required to establish the distribution profile, as long as the movement is diffusion-dominated. Tritium may also be used for tracing experiments, since it apparently may be readily extracted by vacuum distillation and measured accurately by radioactive counting.

Speciation studies are the most affected by the analytical problems that we have pointed out. Here, the whole task of characterising the water content and composition relics on the absence of a clear definition of which part of the cations and anions belongs to the adsorbed and strongly bound clay surface, and which part belongs to the bulk solution. That problem has been neglected so far or treated in terms of total water content. Without that definition, the question would turn up to which total suction needs to be applied to the sample in order to extract even more water, that would actually be no porewater at all. As a consequence, future investigations should aim at a better understanding of the fundamental properties of the clay/water system: thermodynamics of pore-confined water is still a critical issue.

There is a considerable amount of information to be gained by coupling the mechanical behaviour of the system with its mineralogical and chemical characteristics. Macroscopic properties, such as swelling and mechanical strength, depend on the water content, but also on the type of saturating cation on the clay surface and, consequently, on the porewater composition. The establishment of water adsorption-desorption isotherms shows the importance of hysteresis phenomena and may help in the comprehension of the disturbance induced by water extraction.

A rigorous physical study on the size and distribution of pores is needed. The use of different analytical techniques is suggested to define the different porosities (Pearson, in press). Attention should be paid to use relatively "soft" techniques, trying to avoid pore-size modifications related to the force applied to have the fluid permeate through the system. When possible, microscopic observations coupled with the statistic processing of data should be used to evaluate three-dimensionally the rock fabric and the anisotropy of the pore shape. The *in-situ* use of tracers should provide additional and complementary information.

151

Finally, water-content studies should be performed to evaluate the amount of free and bound water. That should be achieved using NMR, IRS and DRS. Those techniques have the advantage to disturb less the clay sample. In addition, recent technical developments allow the use of those techniques directly inside explorative boreholes, bypassing all the artefacts related to sample collection and preservation

Some new and promising water-extraction techniques have been proposed recently for argillaceous materials. Those are the direct equilibration for isotopic analysis (see Chapter II § 3.5.2) and the radial-diffusion method for conservative and non-reactive tracers (mainly deuterium and chloride) (see Chapter II § 3.6.4). Those techniques still need validation on a variety of clay environments to test their applicability to different mixtures of clay minerals and different salinities of the interstitial solutions.

Other topics where further investigation is needed are highlighted in the following paragraphs, and, if possible, the way to proceed to obtain the information.

4.1 Chemical studies

Considering the limited number of reliable tracers available in the relevant clay environments, it may be worth trying to understand better the behaviour of chlorine during extraction by leaching. We have stressed that reproducible results may be obtained for high chloride-content porewaters, while in low-chlorine environments, difficulties appear during extraction. In those cases there seems to be a dependence on the used solid/liquid ratio (see Chapter III § 1.3). That phenomenon may be due to anion adsorption on particular sites under given leaching conditions (pH and salinity of the leaching solution), or be related to porosity-access difficulties. A better understanding of that issue is crucial for the interpretation of leaching data, especially for modelling purposes, and for the definition of effective and geochemical porosity. It may possibly be accomplished by studying the chloride distribution onto and in the solid phase (adsorption sites, fluid inclusions) through mineral and surface analyses.

Although classical CEC measurements are commonly used to define the ion-exchange capacity of clays, the values obtained display high variability depending on the high-affinity cation or complex used (Van Olphen and Fripiat, 1979). There is still a lack of comprehension of the real meaning of that parameter. We have already stressed the importance of always specifying under which conditions the CEC has been measured. An alternative based on thermodynamics consists in characterising the rock by its ion-exchange isotherms, thus enabling to identify specific sites, each having a defined cation-exchange capacity and a selectivity coefficient for each pair of cations. It is only through a multi-site representation (corresponding to the rock structure) and fully taking into account proton sorption, that current ambiguities may be resolved.

Thermodynamic modelling currently seems a promising indirect approach to porewater composition, but more analytical work is needed to determine reliable exchange constants. Acidity and oxido-reduction levels may be readily predicted, but the robustness of the modelling still has to be proven, especially at varying temperatures. In addition, the models should be validated on the research sites, for example, by allowing the *in-situ* equilibration of injected water with the calculated composition with the host rock (see *e.g.,* the pH equilibration experiment conducted at Mont Terri).

Considering the limitations of the available techniques for water extraction, efforts need to be made towards a better understanding of the water-movement mechanisms within the sample that may influence the composition of the extracted water. That requires not only the analysis of the extracted fluids, but also the careful examination of the solid phase prior and after extraction. In that perspective, the ongoing investigations at CIEMAT within the framework of the FEBEX project are extremely interesting, even if they only involve a synthetic system.

Finally, although a leading work has been performed especially on the Boom clay within different projects, a huge amount of study still has to be performed on organic-matter characterisation and the evaluation of its retention role. Organic-matter behaviour may be examined using two very different approaches: ion exchanger or complexing agent. The CEC of organic matter is normally included in the total rock CEC measurements. Few data are available on the organic matter itself, especially if extracted from natural environments. On the other hand, the complexing properties are also relatively unknown. The comparison between values obtained on untreated clay samples and on purified clay fractions may provide useful information regarding the real buffering capability of the organic matter.

4.2   Isotopic studies

Considering isotopic studies, an improvement of existing techniques and/or the development of new techniques, such as direct equilibration, seem possible. That may be achieved via a close coupling with other techniques such as NMR, IRS and DRS, that allow the understanding of the water-type partitioning. In addition, much information may be gained, by the study of homoionic clays, on the role of specific cations in the isotopic fractionation over different types of water. That may be performed by coupling differential thermal analysis with isotopic measurements over the different collected water fractions.

A lack of experimental data on the pressure effect over the isotopic composition of waters extracted by squeezing does not allow a validation of that technique for isotopic analysis. That topic may be further investigated, as that technique, at low pressures, is supposed to affect only free porewater.

4.3   Guidelines for benchmark experiments

At this stage, very few benchmark experiments may be planned. Our suggestions point towards a better integration of the techniques and the knowledge of the fundamental properties of water, organic matter and clay interactions. The scientific community is currently very much aware of the problems involved in water extraction and of the significance of the information to be derived from it.

An intercomparison over analytical techniques on clays has already been organised by the NEA (Van Olphen and Fripiat, 1979) and a second one was launched by the IAEA on isotope analysis (Walker *et al.*, 1994). There is little doubt in our minds that, if those benchmark experiments were to be conducted again today, the results would be anything else but the same.

Nevertheless, a formal organisation of a sample interchange, selecting both plastic and indurated clays with varying water and salinity contents, may be useful. That exchange is actually already going on at a confidential level between different analytical laboratories. That would allow, as far as isotopic techniques are concerned at least, to obtain a clear picture of the extraction technique variability *versus* the sample variability. Teams involved should already have some knowledge of the problem with water extraction from clay-rich environments in order to avoid a large spread of data. The intercomparison may focus as well on the determination of the amount of chloride in porewater, on the determination of the cation-exchange properties (or alternatively, on cation occupancies) and on the isotopic composition of porewater. In any case, it is worth stressing that, for the time being, intercomparisons must only be considered as one of the many tools to understanding better the phenomenology of each analytical method. No intercomparison exercise will provide reliable conclusions as long as no fundamental knowledge exists to assess the accuracy of any method.

GLOSSARY AND FREQUENTLY USED ABBREVIATIONS

absorption[1]   The penetration of a substance into the inner structure of another, as opposed to *adsorption,* where one substance is attracted to and held on the surface of another.

acidity   Of a species, its ability to release protons in a solution; of a solution the proton activity in it.

activity   A magnitude characterising the chemical reactivity of a species or component, equivalent to the chemical potential.

adsorption[1]   The adhesion of atoms, ions or molecules of a gas or liquid to the surface of another substance, called the "adsorbent".

AEC   Anion-exchange capactiy

aliphatic   Used to describe an organic compound with only single-bound carbon atoms.

alkalinity   Of a solution, the concentration of bases that are titrated by a strong acid at the equivalent point corresponding to carbonic acid (Morel).

aluminol   An OH group bound to surface aluminium atoms in a mineral.

aluminosilicate[3]   A silicate in which aluminium substitutes for silicon in the SiO_4 tetrahedra.

amino acid[1]   An organic acid containing both a basic amino group (NH_2) and an acidic carboxyl group (COOH).

ANDRA   National Radioactive Waste Management Agency, France

anion   An ion having a negative charge.

APT[TM]   Accelerator porosity tool (Schlumberger)

aromatic   Used to describe an organic compound having unsaturated six-carbon rings with three delocalised double bonds.

auger[3]   A screw-like boring tool (resembling a carpenter's auger bit but much larger), usually motor driven, designed for clay, soils and other relatively-unconsolidated near-surface materials.

azeotrope[1]   A liquid mixture of two or more substances behaving like a single substance in the sense that the vapour produced by the partial evaporation of liquid has the same composition as the liquid.

bentonite[3]   (a) A soft, plastic, porous, light coloured rock composed essentially of clay minerals of the montmorillonite (smectite) group plus colloidal silica. The rock commonly has the ability to absorb large quantities of water and to increase in volume by about 8 times. The term was first used by Knight (1898) for the argillaceous Cretaceous deposits occurring in the Benton formation in eastern Wyoming.

	(b) A commercial term applied to any of the numerous variously coloured clay deposits containing montmorillonite (smectite) as the essential mineral and presenting a very large total surface area, characterised either by the ability to swell in water or to be stacked and to be activated by acid, and used chiefly (at a concentration of about 3 lb/ft^3 of water) to thicken oil-well drilling muds.
BGS	Britsh Goelogical Survey, United Kingdom
biomarker[3]	An organic compound with a specific structure that may be related to a particular source organism.
bitumen[2]	A native solid or semi-solid hydrocarbon (naphta or asphalt), soluble in carbon disulphate, rich in carbon and hydrogen.
capillarity[3]	The action by which a fluid, such as water, is drawn up (or depressed) in small interstices or tubes as a result of surface tension.
carboxylic acid	An organic molecule having at least one carboxyl group (COOH).
casing[3]	A heavy, large diameter metal pipe, lowered into a borehole during or after drilling and cemented to the sides of the wall. It prevents the sides of the hole from caving, prevents loss of drilling muds or other fluids into porous formations and prevents unwanted fluids from entering the hole.
cation[1]	An ion having a positive charge. Cations in a liquid subjected to an electric potential collect at the negative pole or cathode.
CEA	Atomic Energy Commission, France
CEC	Cation-exchange capacity
CEN/SCK	Nuclear Energy Research Centre, Belgium
chalk[3]	A soft, pure, earthy, fine-textured, usually white to light grey or bluff limestone of marine origin, consisting almost wholly (90-98%) of calcite, formed mainly by shallow water accumulation of calcareous test of floating micro-organisms and calcareous algae, set in a structureless matrix of very finely crystalline calcite. The rock is porous, somewhat friable and only slightly coherent.
colloid[2]	A state of subdivision of matter, comprising either single large molecules or aggregation of smaller molecules. The particles (are) of ultramicroscopic scale. Size range from 10^{-7} to 10^{-5} cm.
CMR™	Combinable magnetic resonance (Schlumberger)
compaction[3]	A reduction in the bulk volume or thickness of, or the pore space within, a body of fine-grained sediments in response to the increasing weight of overlying material that is continuously being deposited or to the pressures resulting from earth movements within the crust. In addition, the process whereby fine-grained sediment is converted to consolidated rock such as clay lithified to a shale.
co-ordination (number)[3]	In crystallography, the number of nearest neighbour ions that surround a given ion in a crystal structure, *e.g.,* four, six or eight.
covalent (bond)[1]	Homopolar. The sharing of electrons by a pair of atoms.
delrin™[2]	A trademark for a linear polyoxymethylene-type acetal resin, having a high strength and solvent resistance. It is mouldable.

desorption[1]	The process of removing an adsorbed material from the solid on which it is adsorbed.
diagenesis[3]	All the chemical, physical and biologic changes undergone by a sediment after its initial deposition and during and after its lithification, excluding surficial alteration and metamorphism. It embraces those processes (such as compaction, cementation, reworking, authigenesis, replacement, crystallisation, leaching, hydration, bacterial action and formation of concretion) that occur under conditions of pressure (up to 1 kbar) and temperature (maximum range of 100° to 300°C) that are normal in the surficial or outer part of the Earth's crust. There is no universally accepted definition of the term and no delimitation (such as the boundary with metamorphism).
dialysis[3]	A method of separating compounds in solution or suspension by their differing rates of diffusion through a semi-permeable membrane, some species not moving through at all, some moving slowly, and other diffusing quite readily. Cf. *osmosis*.
DIC	Dissolved inorganic carbon
dielectric permittivity[2]	(ε_0) the "absolute" permittivity of a dielectric, being the ratio of the electric displacement to the electric-field strength at the same point. Its value for free space is $8.854185 \cdot 10^{-12}$ F m^{-1}. The real component of the complex relative permittivity is an important diagnostic parameter when considering the high- and low-frequency values for zero dielectric loss. It relates to grain size, packing variations and density of the material.
dipole, electric[1]	An assemblage of atoms or subatomic particles having equal electric charges of opposite sign separated by a finite distance; for instance, the hydrogen and chlorine atoms of a hydrogen-chloride molecule.
DOC	Dissolved organic carbon
DRS	Dielectric relaxation spectroscopy
Eh[3]	The potential of a half-cell, measured against the standard hydrogen half-cell. Syn.: oxido-reduction potential.
electric conductivity[3]	A measure of the ease with which a conduction current can be caused to flow through a material under the influence of an applied electric field. It is the reciprocal of resistivity and is measured in siemens per metre.
electrolyte[2]	A substance that dissociates into two or more ions, to some extent, in water. Solutions of electrolytes thus conduct an electric current and may be decomposed by electrolysis.
ENRESA	Spanish National Agency for Radioactive Waste, Spain
ethylenediamine[1]	$NH_2CH_2CH_2NH_2$.
GRS	Company for Reactor Safety, Germany
flocculation[2]	The coagulation of a finely divided precipitate.
fractionation[1]	In general, the separation or isolation of components of a mixture. In distillation, that is done by means of a tower or column in which rising vapour and descending liquid are brought into contact (counter-current flow). Isotopic fractionation changes in isotope ratios occurring for low atomic-number elements that display a large proportional difference in atomic mass between stable isotopes.

fulvic acid[3]   The pigmented organic matter of indefinite composition that remains in solution when an aqueous alkaline extract of soil is acidified by removal of humic acid. It is separated from the fulvic-acid fraction by adsorption on a hydrophilic resin at low pH values.

functional group   A reactive molecular unit that protrudes from a solid adsorbent surface into an aqueous solution (Sposito, 1984).

humic acid[1]   A brown, polymeric constituent of soils, lignite and peat. It is soluble in bases, but insoluble in mineral acids and alcohols. It is not a well-defined compound, but a mixture of polymers containing aromatic and heterocyclic structures, carboxyl groups and nitrogen.

humification[3]   The process whereby the carbon of organic residues is transformed and converted to humic substances through biochemical and abiotic processes.

hydration[2]   The combination with water.

hydrocarbon[1]   An organic compound consisting exclusively of the elements carbon and hydrogen. Derived mainly from petroleum, coal tar and plant sources.

hydrophilic[2]   Lyophilic. Used to describe a substance that readily associates with water.

hydrophobic[2]   Used to describe a substance that does not associate with water.

hysteresis[2]   The lag or retardation of an effect behind its cause, such as (1) the magnetic lag or retention of the magnetic state of iron in a changing magnetic field; (2) the retardation of a chemical system to reach equilibrium.

IAEA   International Atomic Energy Agency

IPSN   Nuclear Protection and Safety Institute, France

IRS   Infrared spectroscopy

isomorphic[3]   Used to described a substance having identical or similar form to another.

isotherm[1]   A constant-temperature line used on climatic maps or in graphs of thermodynamic relations, particularly the graph of pressure/volume relations at constant temperature.

isotope[1]   One of two or more forms or species of an element that have the same atomic number, *i.e.,* the same position in the periodic table, but different masses. The difference in mass is due to the presence of one or more extra neutrons in the nucleus.

kerogen[1]   The organic component of oil shale, it is a bitumen-like solid whose approximate composition is 75-80% carbon, 10% hydrogen, 2.5% nitrogen, 1% sulphur and the balance oxygen. It is a mixture of aliphatic and aromatic compounds of humic and algal origin.

kerosene[2]   A mixture of hydrocarbons, boiling at 150-280°C; the fraction in the distillation of petroleum between gasoline and the oils.

lattice[1]   The structural arrangement of atoms in a crystal.

ligand[1]   A molecule, ion or atom that is attached to the central atom of a co-ordination compound, a chelate or other complex. Ligands are also called "complexing agents".

lignin[1]	A phenylpropane polymer of amorphous structure comprising 17-30% of wood.
LSM	Laboratory strandard marl
lucite[TM2]	A trademark for plastic based on polymerised methyl methacrilate resin.
lyophilisation[1]	A method of dehydration or of separating water from solid materials. The material is first frozen and then placed in a high vacuum in order for the water (ice) to vaporise in the vacuum (sublimes) without melting and for the non-water components to be left behind in an undamaged state.
mole[2]	An SI base unit. The amount of substance of a system containing the same number (*i.e.,* the Avogadro constant) of elementary entities (must specify which; *e.g.,* atoms, molecules, ions, electrons, or other particles) as there are atoms in exactly 0.012 kg of ^{12}C.
monovalent, divalent, trivalent	Cf. *valence*.
mudstone[3]	An indurated mud having the texture and composition of shale, but lacking its fine lamination or fissility; a blocky or massive, fine-grained sedimentary rock in which the proportions of clay and silt are approximately equal.
NAGRA	National Co-operative for the Disposal of Radioactive Waste, Switzerland
NEA	Nuclear Energy Agency
neoprene[2]	Polychloroprene. The generic name for synthetic rubber made by polymerisation of 2-chloro-1,3-butadiene. Neoprene vulcanisates are resistant to oils, chemicals, sunlight, ozone and heat.
NMR	Nuclear magnetic resonance
oedometer[4]	An instrument for testing consolidation of small samples, including the compressibility and consolidation coefficients. The sample is compressed in a cylinder allowing only vertical consolidation; the resulting changes in volume over a specific time are measured.
ONDRAF/ NIRAS	National Organisation for Radioactive Waste and Fissile Materials, Belgium
opt(r)ode	A component of fibre-optical analytical systems that is analogous to an electrode. It comprises a sensor tip and an optical fibre to conduct the light beam carrying the analytical signal.
orbital[1]	The area in space about an atom or molecule in which the probability of finding an electron is greatest.
osmosis[3]	The movement of a solvent through a semi-permeable membrane, that usually separates the solvent and a solution, or a dilute solution from a more concentrated one, until the solutions on both sides of the membrane are equally strong. Cf. *dialysis*.
oxydation	The loss of electrons.

peptide[2]   A component of 2 to 10 amino acids joined through the main (not side) chain by the peptide amine bond, $-C(:O)NH-$. Polypeptides contain 10 to 100 amino-acid residues and resemble peptones and proteins.

perbunam[TM2]   The trademark for a copolymer of butadiene and styrene.

permeability[3]   The property or capacity of a porous rock, sediment or soil, to transmit a fluid. It is a measure of the relative ease of fluid flow under unequal pressure and is a function only of the medium. The customary unit of measurement is the millidarcy.

pH   The symbol for the logarithm of the reciprocal of the hydrogen-ion activity; hence, $pH = \log (1/(H^+))$.

phenol[2]   A compound containing one or more hydroxyl groups attached to an aromatic ring. Polyphenols contain four or more $-OH$ groups.

phtalate[2]   A salt of phtalic acid containing the ion $C_6H_4(COO)_2^{2-}$, used for buffers and standard solutions, and in vacuum pumps.

piezometer[3]   An instrument for measuring pressure heads in a conduit, tank or soil. It usually consists of a small pipe or tube tapped into the side of a container, with its inside end flush with, and perpendicular to, the water face of the container and connected to a manometer pressure gage, or mercury, or water column.

playa[3]   A term used in the southwestern United States for a dry, vegetation-free, flat area at the lowest part of an non-drained desert basin, underlain by stratified clay, silt or sand, and commonly by soluble salts.

polar[1]   Used to describe a molecule in which the positive and negative electric charges are permanently separated, as opposed to non-polar molecules, in which the charges coincide.

polymer   A macromolecule formed by the chemical union of at least two identical units called "monomers". In most cases, the number of monomers is quite large (3,500 for pure cellulose), and often is not precisely known.

porosity[3]   The percentage of the bulk volume of a rock or soil that is occupied by interstices, whether isolated or connected.

potash[3]   a) Potassium carbonate, K_2CO_3; b) A term that is loosely used for potassium oxide, potassium hydroxide or even for potassium in such informal expressions as potash spar; c) An industry term for a group of potassium-bearing salts, that includes potassium chloride, and may also refer to potassium sulphate, nitrate and oxide.

potential[2]   Any stored-up energy capable of performing work. "Chemical potential" is a measure of the tendency of a chemical reaction to take place. Directly related to activity. "Electric potential" is the electromotive force.

precipitate[1]   A set of small particles that have settled out of a liquid or gaseous suspension by gravity, or that results from a chemical reaction.

proton[1]   A fundamental unit of matter having a positive charge and a mass number of 1, equivalent to $1.67 \cdot 10^{-24}$ grams. A proton is identical to a hydrogen ion (H^+).

quasi-crystal   An aggregate structure in coagulated clay minerals (2:1 layer types), in particular smectite, comprising parallel alignments of unit layers.

quick clay[3] A clay that loses all or nearly all its shear strength after being disturbed; a clay that shows no appreciable regain in strength after remoulding.

redox[1] The short form of the term "oxidation-reduction", as in redox reactions, redox conditions, etc. Cf. *oxidation*.

sandstone[3] A medium-grained, clastic sedimentary rock composed of abundant rounded or angular fragments of sand size with or without a fine-grained matrix (silt or clay) and more or less firmly united by cementing material (commonly silica, iron oxide or calcium carbonate). The sand particles are predominantly quarts and the term "sandstone", when used without qualification, indicates a rock containing about 85-90% quartz.

saturation[1] (1) The state in which all available bonds of an atom are attached to other atoms. (2) The state of a solution when it holds the maximum equilibrium quantity of dissolved matter at a given temperature.

selenite[3] The clear, colourless variety of gypsum, occurring (*e.g.*, in clays) in distinct, transparent, monoclinic crystals or in large crystalline masses that easily cleave into broad folia.

S.I. Saturation index

silanol An OH group attached to surface silicon atoms in a mineral

siloxane cavity A surface functional group constituted by a roughly hexagonal cavity formed by six corner-sharing silica tetrahedra. The cavity has a diameter of about 0.26 nm and is bordered by six sets of electron orbitals emanating from the surrounding ring of oxygen atoms (Sposito, 1989).

solid solution[3] A single crystalline phase that may be continuously varied in composition within finite limits without the appearance of an additional phase.

solvation[2] Any stabilising interaction between solute and solvent; if the latter is water, hydrates or hydrated ions are formed, *e.g.*, $M(H_2O)_n$.

sorption[1] A generic, independent of the mechanism, surface phenomenon that may be either absorption or adsorption, or a combination of both. The term is often used when the specific mechanism is not known.

stoichiometric[3] Of a compound or phase, pertaining to the exact proportions of its constituents specified by its chemical formula.

suction[2] The effect of sucking. The operation of drawing fluid into a pipe.

TDS Total dissolved solids

TIC Total inorganic carbon

till[3] A dominantly unsorted and non-stratified drift, generally unconsolidated and deposited directly by and beneath a glacier without subsequent reworking by melt water. It consists in a heterogeneous mixture of clay, silt, sand, gravel and boulders ranging widely in size and shape.

TO Tetrahedral-octahedral layer structure

TOC Total organic carbon

toluene[2] $C_6H_5 \cdot CH_3$. Colourless liquid; density$_{13}$: 0.871; melting point: $-93°C$; boiling point: $111°C$; insoluble in water, but soluble in alcohol or ether.

TOT   Tetrahedral-octahedral-tetrahedral layer structure

tracer[3]   Any substance that is used in a process to trace its course, specifically radioactive material introduced into a chemical, biological or physical reaction.

transmissivity[3]   The rate at which water is transmitted through a unit width of the aquifer under a unit hydraulic gradient. Transmissivity is equal to the product of hydraulic conductivity and aquifer thickness.

tritium[2]   T or ^3H. A radioactive isotope of hydrogen; half-life: 12.5 a, decaying with the emission of β particles.

tuff[3]   A consolidated or cemented volcanic ash.

Urethane[1]   Ethyl carbamate; ethyl urethane; $CO(NH_2)OC_2H_5$. Its structure is typical of the repeating unit in polyurethane resins.

Valence[2]   (1) The capacity of an atom to combine with others in definite proportions. (2) Also applied by analogy, to radicals and atomic groups. The combining capacity of an hydrogen atom is taken as unity, values are integers 1-8.

Van der Waals forces[2]   The weak forces between atoms and molecules, being other than forces due to covalent bonds or ionic attraction.

Vivianite[3]   A mineral $Fe_3^{2+} (H_2O)_8 (PO_4)_2$.

Xylene[2] $C_6H_4Me_2$. Colourless liquid, density: 0.881; melting point: $-28°C$; boiling point: $144°C$; insoluble in water.

Glossary sources:

Definition have been obtained and freely adapted from:
1. Lewis, 1993;
2. Grant and Grant, 1987;
3. Jackson, 1997;
4. Allaby and Allaby, 1990.

REFERENCES

ADAMS, F., BURMESTER, C., HUE, N.V. and LONG, F.L. (1980) "A comparison of column-displacement and centrifuge methods for obtaining soil solutions". *Soil Sci. Soc. Am. J.,* **44**:733-735.

AITEMIN, CIEMAT, CSIC-Zaidin, DM Iberia, ULC, UPC-DIT, UPM and TECNOS (1998) "FEBEX: Full-scale engineered barriers experiment in crystalline host rock. Pre-operational stage Summary Report". ENRESA, Publ. Tecn. 01/98:185.

AJA, S.U., RESENBERG, P.E. and KITTRICK J.A. (1991) "Illite equilibria in solutions: I. Phase relationships in the system $K_2O-Al_2O_3- SiO_2-H_2O$ between 25 and 250 °C". *Geoch. Cosmoch. Acta,* **55**:1353-1364.

ALLABY, A. and ALLABY, M. (1990) *The concise Oxford Dictionary of Earth Sciences.* Oxford University Press, Oxford, 410.

ALLEN, A.J., BASTON, A.H., BOURKE, P.J. and JEFFERIES, N.L. (1988) "Small angle neutron scattering studies of diffusion and permeation through pores in clays". Nirex Safety Studies Rep., NSS/R160:38.

ALLISON, G.B., BARNES, C.J. and HUGHES, M.W. (1983) "The distribution of deuterium and ^{18}O in dry soils: 2. Experimental". *J. Hydrol.,* **64**:377-397.

ALLISON, G.B., COLIN-KACZALA, C., FILLY, A. and FONTES, J.Ch. (1987) "Measurement of isotopic equilibrium between water, water vapour and soil CO_2 in arid zone soils". *J. Hydrol.,* **95**:131-141.

ALLISON, G.B. and HUGHES, H.W. (1983) "The use of natural tracers as indicators of soil water movement in a temperate semi-arid region". *J. Hydrol.,* **60**:157-173.

ANDERSON, D.M. and LOW, P.F. (1958) "The density of water adsorbed on lithium-, sodium-and potassium-bentonite". *Soil Sci. Soc. Amer Proc.,* **22**:99-103.

ANDREUX, F. and MUNIER-LAMY, C. (1994) "Genèse et propriétés des molécules humiques". In: *Pédologie, constituants et propriétés du sol* (Bonneau M. and Soucher B., Eds.), Masson, **2**:109-142.

ANDREWS, J.N., DAVIS, S.N., FABRYKA-MARTIN, J., FONTES, J.C., LEHMANN, B.E., LOOSLI, H.H., MICHELOT, J.-L., MOSER, H., SMITH, B. and WOLF, M. (1989) "The in-situ production of radioisotopes in rock matrices, with particular reference to the Stripa granite". *Geoch. Cosmoch. Acta,* **53**:1803-1815.

ANGELIDIS, T.N. (1997) "Comparison of sediment porewater sampling for specific parameters using two techniques". *Water, air, and soil poll.,* **99**:179-185.

APPELO, C.A.J. (1977) "Chemistry of water expelled from compacting clay layers: a model based on Donnan equilibrium". *Chem. Geol.,* **19**:91-98.

APPELO, C.A.J. (1994) "Cation and proton exchange, pH variations and carbonate reactions in a freshening aquifer". *Water Res. Res.,* **30**/10:2793-2805.

APTE, S.C., GARDNER, M.J. and HUNT, T.E. (1989) "An evaluation of dialysis as a size-based separation method for the study of trace metal speciation in natural waters". *Environ. Techn. Lett.,* **10**:201-210.

AQUILINA, L., BOULÈGUE, J., PINAULT J.-L. and SUREAU J.-F. (1993b) "WELCOM (Well Chemical On-line-Monitoring) II: Chemical interpretation of WELCOM, Balazuc-1 well, Ardèche France". *Scient. Drilling,* **4**:13-22.

AQUILINA, L., BOULÈGUE, J., SUREAU, J.-F., BARIAC, T. and GPF TEAM (1994) "Evolution of interstitial waters along the passive margin of the Southeast Basin of France: WELCOM (Well Chemical On-line Monitoring) applied to Balazuc-1 well (Ardèche)". *Appl. Geoch.,* **9**:657-675.

AQUILINA, L., CÉCILE J.-L., SUREAU J.-F. and DEGRANGES P. (1993a) "WELCOM (Well Chemical On-line-Monitoring) I: Technical and economic aspects". *Scient. Drilling,* **4**:5-12.

AQUILINA, L., DELAY, J. and MERCERON, T. (1995) "Characterisation of interstitial water and ionic exchange in the clay formation through monitoring of drilling fluids". *Proc. of an Int. Workshop on "Hydraulic and hydrochemical characterisation of argillaceous rocks",* U.K., 7-9 June 1994, OECD/NEA, 119-131.

AQUILINA, L., EBERSCHWEILER, C., PERRIN, J. and DEEP GEOLOGY OF FRANCE TEAM (1996) "Comparison of hydrogeochemical logging of drilling fluid during coring with the results from geophysical logging and hydraulic testing. Example of the Morte-Mérie scientific borehole, Ardèche, France, Deep Geology of France Programme". *J. Hydrol.,* **185**:1-21.

ARAGUAS ARAGUAS, L., ROZANSKI, K., GONFIANTINI, R. and LOUVAT, D. (1995) "Isotope effects accompanying vacuum extraction of soil water for stable isotope analyses". *Journ. Hydrol.,* **168**:159-171.

ARANYOSSY, J.-F. and GAYE, C.B. (1992) "La recherche du pic de tritium en zone non saturée profonde sous climat semi-aride pour la mesure de la recharge des nappes: première application au Sahel". *C.R. Acad. Sci. Paris,* 315, II, 637-643.

AZCUE, J.M.A., ROSA, F. and LAWSON, G. (1996) "An improved dialysis sampler for the in situ collection of larger volumes of sediment porewaters". *Environ. Techn.,* **17**:95-100.

BAEYENS, B. and BRADBURY, M.H. (1991) "A physico-chemical characterisation technique for determining the porewater chemistry in argillaceous rocks". NAGRA Tech. Rep., 90-40, 61.

BAEYENS, B. and BRADBURY, M.H. (1994) "Physico-chemical characterisation and calculated in situ porewater chemistries for a low permeability Palfris Marl sample from Wellenberg". NAGRA Tech. Rep., 94-22, 30.

BARNES, C.J. and ALLISON, G.B. (1983) "The distribution of deuterium and ^{18}O in dry soils: 1. Theory". *J. Hydrol.,* **60**:151-156.

BARNES, C.J. and ALLISON, G.B. (1984) "The distribution of deuterium and ^{18}O in dry soils: 3. Theory for non isothermal water movement". *J. Hydrol.,* **74**:119-155.

BARNEYBACK, R.S. Jr. and DIAMOND, S. (1981) "Expression and analysis of pore fluids from hardened cement pastes and mortars". *Cement and Concrete Research,* **11**:279-285.

BATH, A.H. and EDMUNDS, W.M. (1981) "Identification of connate water in interstitial solution of chalk sediment". *Geoch. Cosmoch. Acta,* **45**:1449-1461.

BATH, A.H., ENTWISLE, D., ROSS, C.A.M., CAVE, M.R., FALCK, W.E., FRY, M., REEDER, S., GREEN, K.A., McEWEN, T.J. and DARLING, W.G. (1988) "Geochemistry of porewaters in mudrock sequences: evidence for groundwater and solute movements. *Proc. of the Int. Symp. on "Hydrogeology and safety of radioactive and industrial hazardous waste disposal"*, IAH, BRGM Doc. 160, 1, 87-97.

BATH, A.H., ROSS, C.A.M., ENTWISLE, D., CAVE, M.R., FRY, M.B., GREEN, K.A. and REEDER, S. (1989a) "Hydrochemistry of porewaters from Lower Lias Siltstones and limestones at the Fulbeck site". B.G.S. Fluid Processes Group Tech. Rep., WE/89/27, 44.

BATH, A.H., ROSS, C.A.M., ENTWISLE, D., CAVE, M.R., GREEN, K.A., REEDER, S. and FRY, M.B. (1989b) "Hydrochemistry of porewaters from London Clay Lower London Tertiaries and Chalk at the Bradwell Site". B.G.S. Fluid Processes Group Tech. Rep., WE/89/26, 131.

BATLEY, G.E. and GILES, M.S. (1979) "Solvent displacement of sediment interstitial waters before trace metal analysis". *Wat. Res.,* **13**:879-886.

BAYLEY S.W. (Ed.) (1988) "Hydrous phyllosilicates (exclusive of micas)", *Reviews in Mineralogy,* 19:725.

BEAUCAIRE, C., PITSCH, H. and BOURSAT, C. (1998) "Modelling of redox conditions and control of trace elements in clayey groundwater". *Water Rock Interaction WRI 9,* Taupo, New Zealand, 30 March-3 April 1998 (Arehart, G.B. and Huston, J.R., Eds.), 141-144.

BEAUCAIRE, C., PITSCH, H., TOULHOAT, P., MOTELLIER, S. and LOUVAT, D. "Regional fluid characterisation and modelling of water-rock equilibria in the Boom clay formation and in the Rupelian aquifer at Mol, Belgium". *Applied Geochemistry.* (in press)

BEAUCAIRE, C., TOULHOAT, P. and PITSCH, H. (1995) "Chemical characterisation and modelling of the interstitial fluid in the Boom clay formation". *Proc. of the VIII Int. Symp. on Water-Rock Interaction WRI-8* (Kharala, Y.K., Chudae,v O.V. Eds) Vladivostok, Russia, 15-19 August 1995, Balkema, 789-782.

BEAUFAYS, R., BLOOMAERT, W., BRONDERS, P., DE CANNIÈRE, P., DELMARMOL, P., HENRION, P., MONSECOUR, M., PATYN, J. and PUT, M. (1994) "Characterisation of the Boom clay and its multilayered hydrogeological environment. Final Report". EUR 14961, 340.

BENDER, M., MARTIN, W., HESS, J., SAYLES, F., BALL, L. and LAMBERT, C. (1987) "A whole-core squeezer for interfacial porewater sampling". *Limnol. Oceanogr.,* **32**/6:1214-1225.

BEN-RHAIEM, H., PONS, C.H. and TESSIER, D. (1987) "Factors affecting the macrostructure of smectites. Role of cation and history of applied stresses". *Proc. of the Int. Clay Conf. 1985* (Schultz, L.G., Van Olphen, H. and Mumpton, F.A,. Eds.), Clay Minerals Soc. Bloomington, Indiana, 292-297.

BENZEL, W.M. and GRAF, D.L. (1984) "Studies of smectite membrane behaviour: importance of layer thickness and fabrics in experiments at 20°C". *Geoch. Cosmoch. Acta,* **48**:1769-1778.

BERNER, R.A., SCOTT, M.R. and THOMLINSON, C. (1970) "Carbonate alkalinity in the porewaters of anoxic marine sediments". *Limn. Oceanog.,* **15**:544-549.

BISCHOFF, J.L., GREER, R.E. and LUISTRO, A.O. (1970) "Composition of interstitial waters of marine sediments: temperature of squeezing effect". *Science,* **167**:1245-1246.

BLACKWELL, P.A., REEDER, S., CAVE, M.R., ENTWISLE, D.C., TRICK, J.K., HUGHES, C.D., GREEN, K.A. and WRAGG, J. (1995a) "Clay porewater and gas analysis preliminary investigation programme at Haute Marne, France". ANDRA, BRP0BGS 95.001

BLACKWELL, P.A., REEDER, S., CAVE, M.R., ENTWISLE, D.C., TRICK, J.K., HUGHES, C.D., GREEN, K.A. and WRAGG, J. (1995b) "Clay porewater and gas analysis preliminary investigation programme at Meuse, France". ANDRA, BRP0BGS 95.002

BOEK, E.S., COVENEY, P.V. and SKIPPER, N.T. (1995) "Molecular modeling of clay hydration: a study of hysteresis loops in the swelling curves of sodium montmorillonites". *Langmuir,* **11**:4629-4631.

BOISSON, J.-Y., CABRERA, J and DE WINDT, L. (1998) "Étude des écoulements dans un massif argileux: laboratoire souterrain de Tournemire". CEC EUR, 18338 FR, 300.

BOLT, G.H. (Ed.) (1982) *Soil chemistry.* Developments in soil science, 5A-5B, Elsevier, Amsterdam.

BÖTTCHER, G., BRUMSACKS, H.J., HEINRICHS, H. and POHLMANN, M. (1997) "A new high-pressure squeezing technique for pore-fluid extraction from terrestrial soils". *Water, Air and Soil Poll.,* **94**:289-296.

BRADBURY, M.H. and BAEYENS, B. (1997) "Derivation of in situ Opalinus clay porewater compositions from experimental and geochemical modelling studies". NAGRA Tech. Rep. NTB 97-07:50.

BRADBURY, M.H. and BAEYENS, B. (1998) "A physicochemical characterisation and geochemical modelling approach for determining porewater chemistries in argillaceous rocks". *Geoch. Cosmoch. Acta,* **62**/5:783-795.

BRADBURY, M.H., BAEYENS, B. and ALEXANDER, W.R. (1990) "Experimental proposals for procedures to investigate the water chemistry, sorption and transport properties of marl". NAGRA Tech. Rep., 90-16:67.

BRADBURY, M.H., BAEYENS, B., PEARSON, F.J and BERNER, U. (1997) "Addendum to derivation of in situ Opalinus clay porewater compositions from experimental and geochemical modelling studies". NAGRA, Tech. Rep. NTB 97-07/ADD, 13.

BRAY, J.T., BRICKER, O.P. and TROUP, B.N. (1973) "Phosphate in interstitial waters of anoxic sediments: oxidation effects during sampling procedure". *Science,* **180**:1362-1364.

BRIGGS, L.J. and McLANE, J.W. (1907) "The moisture equivalent of soils". U.S. Dept. Agriculture, Bureau of Soils Bull., **45**:1-23.

BRIGHTMAN, M.A., BATH, A.H., CAVE, M.R. and DARLING W.G. (1985) "Pore fluids from the argillaceous rocks of the Harwell region". British Geological Survey Rep., FLPU 85-6:77.

BRINDLEY, G.W. and BROWN, G. (1980) "Crystal structures of clay minerals and their X-ray identification". Mineralogical Society, London, 495.

BROWN, G., NEWMAN, A.C.D., RAYNER, J.H. and WEIR, A.H. (1978) "The structure and chemistry of soil clay minerals". In: *The chemistry of soil constituents* (Greenland D.J. and Hayes M.B.H. Eds.), Wiley, Chichester, 29-178.

BUFFLE, J. (1984) "Natural organic matter and metal-organic interactions in aquatic systems". *Metal Ions Biol. Systems,* **18**:165.

BUFFLE, J. (1988) *Complexation reactions in aquatic systems: an analytical approach.* Ellis Horwood series in Analytical Chemistry, Wiley.

BURGESS, J. (1988) *Ions in solution: basic principles of chemical interactions.* Ellis Horwood Publisher, 192.

CAMERON, F.K. (1911) *The soil solution: the nutrient medium for plants.* Easton, Pa., cited in Kriukov and Manheim, 1982.

CARIGNAN, R. (1984) "Interstitial water sampling by dialysis. Methodological notes". *Limnol. Oceanogr., 29*/3:667-670.

CASES, J. and FRANÇOIS, M. (1982) "Étude des propriétées thermodynamiques de l'eau au voisinage des interfaces". *Agronomie, 2*/10:931-938.

CAVE, M.R. and REEDER, S. (1995) "Reconstruction on in situ porewater compositions obtained by aqueous leaching of drill core: an evaluation using multivariate statistical deconvolution". *The Analyst, 120*/5:1341-1351.

CAVE, M.R., REEDER, S. and ARTAZ, J. (1995) "Preliminary evaluation of physical and statistical methods to determine the chemical composition of clay porewaters". In: *Hydraulic and hydrochemical characterisation of argillaceous rocks,* OECD/NEA, Paris, 235-246.

CAVE, M.R., REEDER, S., ENTWISLE, D.C., BLACKWELL P.A., TRICK J.K., WRAGG J. and BURDEN S.R. (1997) "Chemical characterisation of squeezed porewaters and aqueous leachates in Shales from the Tournemire tunnel, France". Brit. Geol. Surv. Techn. Rep., WI/97/6C, 26.

CAVE, M.R., REEDER, S. and METCALFE, R. (1994) "Chemical characterisation of core porewaters for deep borehole investigations at Sellafield, Cumbra". *Mineralogical Magazine, 58A:*158-159.

CHAPMAN, D.L. (1913) "A contribution to the theory of electrocapillarity". *Phil. Mag., 25:*475-481.

CHARLES, R.J., COOK, A.J. and ROSS, C.A.M. (1986) "Solute transport processes in a saturated clay". Brit. Geol. Surv. Fluid Processes Group Report, WE/FL/86/11:27.

CHARLTON, S.R., MACKLIN, C.L. and PARKHURST, D.L. (1997) "PHREEQCI: a graphical user interface for the geochemical computer program PHREEQC". US Geol. Serv. Water-Resources Investigations, 97-4222:9.

CHERRY, J.A., DESAULNIERS, D.E., FRIND, E.O., FRITZ, P., GEVAERT, D.M., GILLHAM, R.W. and LELIEVRE, B. (1979) "Hydrogeologic properties and porewater origin and age: clayey till and clay in South Central Canada". *Proc. of the Workshop on "Low flow, low permeability measurements in largely impermeable rocks"*, Paris, 19-21/03/1979, NEA, 31-46.

CHILINGARIAN, G.V., SAWABINI, C.T. and RIEKE, H.H. (1973) "Effect of compactation on chemistry of solutions expelled from the montmorillonite clay saturated in seawater". *Sedimentology, 20*:391-398.

COPLEN, T.B. and HANSHAW, B.B. (1973) "Ultrafiltration by a compacted clay membrane I. Oxygen and Hydrogen isotopic fractionation". *Geoch. Cosmoch. Acta, 37*:2295-2310.

CONWAY, B.E. (1981) *Ionic hydration in Chemistry and Biophysics.* Elsevier, 741.

CRAIG, H. (1957) "Isotopic standards for carbon and oxygen and correction factors for mass spectrometric analysis of carbon dioxide". *Geoch. Cosmoch. Acta, 12*:133-149.

CROUSSARD, P., WIN, P., HAGOOD, M., LEWIS, R. and STROBEL, J. (1998) "Applications of high resolution borehole geophysical logging to argillaceous forrmations pertinent to nuclear repository siting". In: "Detection of sedimentary and structural heterogeneitites and discontinuities with argillaceous formations". NEA Clay Club Topical Session, Brussels, Belgium, 03/06/98, NEA/SEDE/DOC(98)4,29-32.

CUEVAS, J., VILLAR, M.V., FERNÁNDEZ, A.M., GOMEZ, P. and MARTIN, P.L. (1997) "Porewaters extracted from compacted bentonite subjected to simultaneous heating and hydration". *Appl. Geochem.,* **12**:473-481.

DAHLGREN, R.A., PERCIVAL, H.J. and PARFITT, R.L. (1997) "Carbon dioxide degassing effects on soil solutions collected by centrifugation". *Soil Science,* **162**/9:648-655.

D'ANS, J. (1968) "Der Ubergangspunkt Gips-Anhydrite". Kali Steinsaltz Heft, 3, 109.

DARLING, W.G., METCALFE, R., CRAWFORD, M.B. and HOOKER, P.J. (1995) "The estimation of solute and groundwater residence times in different lithologies: a review of applications of methods". Brit. Geol. Serv. Fluid Processes Group Tech. Rep., WE/94/43:120.

DAVIES, B.E. and DAVIES, R.J. (1963) "A simple centrifugation method for obtaining small samples of soil solution". *Nature,* **198**/4876:216-217.

DAVIDSON, G.R., HARDIN, E.L. and BASSETT, R.L. (1995) "Extraction of ^{14}C from porewater in unsaturated rock using vacuum distillation". *Radiocarbon,* **37**:861-874.

DAVIS J.A. and HAYES K.F. (Eds.) (1985) "Geochemical processes at mineral sufaces". Am. Chem. Soc. Symp. Ser., 323, The American Chemical Society.

DAVISON, W., GRIME, G.W., MORGAN, J.A.W. and CLARKE, K. (1991) "Distribution of dissolved iron in sediment porewaters at submillimetre resolution". *Nature,* **352**:323-325.

DECARREAU, A. (Ed.) (1990) "Matériaux argileux. Structure, propriétés et applications". Société Française de Minéralogie et de Cristallographie, Groupe français des argiles, 586.

DE HAAN, F.A.M. (1965) "The interaction of certain inorganic anions with clays and soils". Agricultural Research Reports 655, Centre for Agricultural Publications and Documentation, Wageningen, 167.

DE LA CALLE, C., SUQUET, H. and PEZERAT, H. (1985) "Vermiculites hydratées à une couche". *Clay Min.,* **20**:221-230.

DE LANGE, G.J. (1986) "Early diagenetic reactions in interbedded pelagic and turbiditic sediments in the Nares Abyssal Plain (Western North Atlantic): consequences for the composition of sediment and interstitial water". *Geoch. Cosmoch. Acta,* **50**:2543- 2561.

DE LANGE, G.J. (1992) "Shipboard routine and pressure-filtration system for porewater extraction from suboxic environments". *Mar. Geol.,* **109**:77-81.

DE LANGE, G.J., CRANSTON, R.E., HYDES, D.H. and BOUST, D. (1992) "Extraction of porewater from marine sediments: A review of possible artifacts with pertinent examples from the North Atlantic". *Mar. Geol.,* **109**:53-76.

DEMIR I. (1988) "Studies of smectite membrane behaviour: electrokinetic, osmotic and isotopic fractionation at elevated pressures". *Geoch. Cosmoch. Acta,* **52**:727-737.

DESAULNIERS, D.E., CHERRY, J.A. and FRITZ, P. (1981) "Origin, age and movement of porewater in the argillaceous Quaternary deposits at four sites in southwestern Ontario". *J. Hydrol.,* **50**:231-257.

DEVINE, S.B., FERREL, R.E. and BILLINGS, G.K. (1973) "The significance of ion exchange to interstitial solutions in clayey sediments". *Chem. Geol.,* **12**:219-228.

DEVOL-BROWN, I., PITSCH, H., LY, S., MEIER, P. and BEAUCAIRE, C. (1998) " Séparation et purification des fractions organique et argileuse d'une roche alumino-silicatée ". *Réunion spécialisée ASF-SGF Argile : sédimentologie, diagenèse, environnement,* Lille, France.

DEWAR, W.A. and McDONALD, P. (1961) "Determination of dry matter in silage by distillation with toluene". *J. Sci. Food Agric.,* **12**:790-795.

DE WINDT, L., CABRERA, J. and BOISSON, J.-Y. (1998a) "Hydrochemistry in an indurated argillaceous formation (Tournemire tunnel site, France)". *Water Rock Interaction WRI 9, Taupo, New Zealand, 30 March –3 April 1998* (Arehart, G.B. and Huston, J.R., Eds.), 145-148.

DE WINDT, L., CABRERA, J. and BOISSON, J.-Y. (1998b) "Radioactive waste containement in indurated claystones: comparison between the chemical containment properties of matrix and fractures". Geological Society of London, Spec. Publ. on Chemical Containment of Waste in the Geosphere, in press, 20.

DE WIT, R. (1995) "Measurements of sedimentary gradients of porewater species, by use of microelectrodes. Calculation of microbial metabolic processes in the sediments". *Oceanis,* **21**/1:287-297.

DIERCKX, A., MAES, A., HENRION, P. and DE CANNIÈRE, P. (1996) "Potentiometric titrations of humic substances extracted from Boom clay". *Proc. 8th Int. Symp Humic Substances Soc.,* Wroclaw, Poland.

DIXON, J.B. and WEEDS, B. (1977) *Minerals in soil environments.* Soil Sci Soc. Am., Madison, Wisconsin.

EDMUNDS, W.M. and BATH, A.H. (1976) "Centrifuge extraction and chemical analysis of interstitial waters". *Environ. Sci. Technol.,* **10**/5:467-472.

EDMUNDS, W.M., KINNINBURGH, D.G. and MOSS, P.D. (1992) "Trace metals in interstitial waters from sandstones: acidic inputs to shallow groundwaters". *Env. Poll.,* **77**:129-141.

EGER, I., CRUZ-CUMPLIDO, M.I. and FRIPIAT, J.-J. (1979) "Quelques données sur la capacité calorifique et les propriétés de l'eau dans différents systèmes poreux". *Clay Min.,* **14**:161-172.

EMERSON, S., JAHNKE, R., BENDER, M., FROELICH, P., KLINKHAMMER, G., BOWSER, C. and SETLOCK, G. (1980) "Early diagenesis in sediments from the Eastern equatorial pacific. I. Porewater nutrient and carbonate results". *Earth Planet. Sci. Lett.,* **49**:57-80.

ENGELHARDT W. and GAIDA K.H. (1963) "Concentration changes of pore solutions during the compactation of clay sediments". *J. Sedim. Petr.,* **33**/4:919-930.

ENTWISLE, D.C. and REEDER, S. (1993) "New apparatus for pore fluid extraction from mudrocks for geochemical analysis". In: *Geochemistry of Clay-Pore fluid interactions* (Manning, D.A.C., Hall, P.L. and Hugues, C.R,. Eds.), The Mineralogical Society Series **84**:365-388.

EPSTEIN, S. and MAYEDA, T.K. (1953) "Variation in ^{18}O content of waters from natural sources". *Geoch. Cosmoch. Acta,* 4:213-224.

FALCK, W.E., BATH, A.H. and HOOKER, P.J. (1990) "Long term solute migration profiles in clay sequences". Zeitschrift der Deutschen Geologischen Gesellschaft, 141, 415-426.

FANNING, K.A. and PILSON, M.E.Q. (1971) "Interstitial silica and pH in marine sediments: some effects of sampling procedures". *Science,* **173**:1228-1231.

FELBECK, G.T. (1971) "Chemical and biological characterization of humic matter". In: Soil biochemistry (McLaren, A.D.R. and Skujins, J., Eds.), 2, Marcel Dekker, New York, 36-59.

FERNÁNDEZ, A.M., CUEVAS, J. and MARTÍN, P.L. (1996) "Equipo para la extracción y determinación de parámetros químicos en aguas intersticiales de bentonitas obtenidas por compactación". FEBEX Project, Int. Rep. CIEMAT/IMA/54A40/5/96.

FERNÁNDEZ, A.M., *et al.* (in prep.) "Study of the porewaters of the FEBEX bentonite". FEBEX Project.

FONTANIVE, A., GRAGANI, R., MIGNUZZI, C. and SPAT G. (1985) "Determinazione delle caratteristiche geochimiche delle acque interstiziali di argille plio-quaternarie italiane". ENEA, Rapp. Fin., 151-81-7 WASI, 42.

FONTANIVE, A., GRAGNANI, R., MIGNUZZI, C. and SPAT G. (1993) "Chemical composition of porewaters in Italian Plio-Pleistocene clayey formations". In: *Geochemistry of Clay-Pore fluid interactions* (Manning D.A.C., Hall P.L. and Hugues C.R. Eds.), The Mineralogical Society Series **84**:389-411.

FÖRSTER, A., SCHÖTTER, J., MERRIAM, D.F. and BLACKWELL, D.D. (1997) "Application of optical-fiber temperature logging. An example in sedimentary environment". *Geophysics,* **62**/4:1107-1113.

FRANCE-LANORD, C. (1997) "Bilan isotopique de l'oxygène et de l'hydrogène de l'eau dans des formations argileuses du forage Est 104 de l'ANDRA". ANDRA, BRP O CRP 97-001

FRANCE-LANORD, C. and SHEPPARD, S.M.F. (1992) "Hydrogen isotope composition of porewaters and interlayer water in sediments from the Central Western Pacific". Proc. ODP, *Sci. Res.* **129**:295-302.

FRANKS, F. (1985) "Water". Royal Soc. Chem. London, 96.

FRANT, M.S. (1997) "Where did ion selective electrodes come from? The story of their development and commercialization". *J. Chem. Education,* **74**/2:159-166.

FREYSSINET, P. and DEGRANGES, P. (1989) "Étude préliminaire concernant le conditionnement et l'extraction des eaux interstitielles des échantillons du forage A901". BRGM Rep., SGN 391 STO.

FRIEDMAN, H.L. (1985) "Hydration complexes: some firm results and some pressing questions". *Chemica Scripta,* **25**:42-48.

FRIEDMAN, I. and SMITH, R.L. (1958) "The deuterium content of water in some volcanic glasses". *Geoch. Cosmoch. Acta,* **15**:218-228.

FRIPIAT, J.J., LETELLIER and M., LEVITZ, P. (1984) "Interaction of water with clay surfaces". Phil. Trans. R. Soc. Lond., A 331, 287-299.

FRITZ, B. (1975) "Étude thermodynamique et simulation des réactions entre minéraux et solutions. Application à la géochimie des altérations et des eaux continentales". Sciences Géologiques, Mém. 41, Un. Louis Pasteur, Strasbourg, 152.

FRITZ, S.J. and EADY, C.D. (1985) "Hyperfiltration-induced precipitation of calcite". *Geoch. Cosmoch. Acta,* **49**:761-768.

GARRELS, R. M. (1984) "Montmorillonite/illite stability diagrams". *Clay and Clay Min.,* **32**/3:161-163.

GAUCHER, E.C., CLAUDE, L. and PITSCH, H., LY, J. (1998) "Influence of temperature on the sorption isotherm of potassium on a montmorillonite". *Water Rock Interaction WRI 9, Taupo, New Zealand, 30 March-3 April 1998* (Arehart, G.B. and Huston, J.R., Eds.), 931-934.

GEDROIZ, K.K. (1906) "On the changeability of composition of soil solution". *Zhurnal Opyt. Agron.,* **7**/5:521-545. (in Russian, cited in Kriukov and Manheim, 1982)

GILLMAN, G.P. (1979) "A proposed method for the measurement of exchange properties in highly weathered soils". *Australian J. Soil Res.,* **17**:129-139.

GODFREY, J.D. (1962) "The deuterium content of hydrous minerals from the East-Central Sierra Nevada and Yosemite National Park". *Geoch. Cosmoch. Acta,* **26**:1215-1245.

GONFIANTINI, R. (1986) "Environmental isotopes in lake studies". In: "Handbook of environmental isotope geochemistry" (Fritz P. and Fontes J.C. Eds.), **2**:113-168, Elsevier, Amsterdam.

GONFIANTINI, R. and FONTES, J.C. (1963) "Oxygen isotopic fractionation in the water of crystallization of gypsum". *Nature,* **200**:644-646.

GORGEON, L. (1994) *Contribution à la modélisation physico-chimique de la retention de radioéléments à vie longue par des matériaux argileux.* Thesis, Université de Paris 6, 201.

GOUVEA DA SILVA, R. (1980) *Migration des sels et des isotopes lourds à travers des colonnes de sédiments non saturés sous climat semi-aride.* Thesis, Université de Paris-Sud.

GOUY, G. (1910) "Sur la constitution de la charge électrique à la surface d'un électrolyte". *J. Phys.,* **9**:457-468.

GRAF, D.L. (1982) "Chemical osmosis, reverse chemical osmosis, and the origin of subsurface brines". *Geoch. Cosmoch. Acta,* **46**:1431-1448.

GRAHAME, D.C. (1947) "The electric double layer and the theory of electrocapillarity". *Chem. Rev.,* **41**:441

GRANT, R. and GRANT, C. (1987) *Grant and Hackh's Chemical Dictionary.* McGraw-Hill Book Company, V Ed., 641.

GRIFFAULT, L., MERCERON, T., MOSSMAN, J.R., NEERDAEL, B., DE CANNIÈRE, P., BEAUCAIRE, C., DAUMAS, S., BIANCHI, A. and CHRISTEN, R. (1996) "Acquisition et régulation de la chimie des eaux en milieu argileux pour le projet de stockage de déchets radioactifs en formation géologique. Projet Archimede-Argile". CEC, EUR 17454, 176.

GRIMAUD, D., BEAUCAIRE and C., MICHARD, G. (1990) "Modelling the evolution of ground waters in a granite system at low temperature: the Stripa ground waters, Sweden". *Appl. Geoch.,* **5**:515-525.

GUGGENHEIM, S. and MARTIN, R.T. (1995) "Definition of clay and clay mineral: joint report of the AIPEA and CMS nomenclature committees". *Clay Min.,* **30**:257-259.

GÜVEN, N. (1988) "Smectites". In: "Hydrous phyllosilicates (exclusive of micas)", Reviews in Mineralogy (Bailey S.W. Ed.), **19**:497-559.

GÜVEN, N. (1992) "Molecular aspects of clay-water interactions. In: Clay water interface and its rheological implications". CMS Workshop Lectures vol. 4 (Güven N. and Pollastro R.M. Eds.), The Clay Minerals Society, 1-80.

GVIRTZMAN, H. and GORELICK, S.M. (1991) "Dispersion and advection in unsaturated porous media enhanced by anion exclusion". *Nature, 352*:793-795.

HALES, B. and EMERSON, S. (1996) "Calcite dissolution in sediments of the Ontong-Java Plateau: in situ measurements of porewater O_2 and pH". *Global Biogeochem. Cycles,* **10**/3:527-541.

HALES, B., EMERSON, S. and ARCHER, D. (1994) "Respiration and dissolution in the sediments of the western North Atlantic: estimates from models of in situ microelectrode measurements of porewater oxygen and pH". *Deep-Sea Res.* I, **41**/4:695-719.

HALL, P.L. (1993) "Mechanisms of overpressuring: an overview". In: Geochemistry of Clay-Pore fluid interactions (Manning D.A.C., Hall P.L. and Hugues C.R. Eds.) The Mineralogical Society Series **84**:265-315.

HANSHAW, B.B. and COPLEN, T.B. (1973) "Ultrafiltration by a compacted clay membrane. II, Sodium ion exclusion at various ionic strenghts". *Geoch. Cosmoch. Acta,* **37**:2311-2327.

HARDCASTLE, J.H. and MITCHELL, J.K. (1974) "Electrolyte concentration-permeability relationships in sodium illite-silt mixtures". *Clays and Clay Min.*, **22**:143-154.

HARTMANN, M. (1965) "An apparatus for the recovery of interstitial water from Recent sediments". *Deep-Sea Res.,* **12**:225-226.

HAYES, M.H.B. (1985) "Extraction of humic substances from soils". In: *Humic substances in soil, sediment and water,* John Wiley, New York, 329-362.

HAYES, K.F. and LECKIE, J.O. (1987) "Modeling ionic strenght effects on cation adsorption at hydrous oxide/solution interfaces". *J. Coll. Interf. Sci.,* **115**:564-572.

HEINRICHS, H., BÖTTCHER, G., BRUMSACK, H.J. and POHLMANN, M. (1996) "Squeezed soil-pore solutions. A comparison to lysimeter samples and percolation experiments". *Water, Air and Soil Poll.,* **89**:189-204.

HENDRICKS, S.B., NELSON, R.A. and ALEXANDER, L.T. (1940) "Hydration mechanism of the clay mineral montmorillonite saturated with different cations". *J. Amer. Chem. Soc.,* **62**:1457-1464.

HENDRY, M.J. (1983) "Groundwater recharge through a heavy textured soil". *J. Hydrol.,* **63**:201-209.

HESSLEIN, R.H. (1976) "An in situ sampler for close interval porewater studies". *Limnol. Oceanogr.,* **21**:912-914.

HOCHELLA, M.F. and WHITE, A.F. (Eds.) (1990) "Mineral-water interface geochemistry". *Reviews in Mineralogy,* 23, The Mineralogical Society of America, 603.

HORSEMAN, S.T., HIGGO, J.J.W., ALEXANDER, J. and HARRINGTON, J.F. (1996) "Water, gas and solute movement through argillaceous media". OECD/NEA, CC-96/1:290.

HOWER, J., ESLINGER, E.V., HOWER, M.E. and PERRY, E.A. (1976) "Mechanism of burial metamorphism of argillaceous sediment. I. Mineralogical and chemical evidences". *Geol. Soc. Am. Bull.,* **87**:725-736.

HULBERT, M.H. and BRINDLE, M.P. (1975) "Effects of sample handling on the composition of marine sedimentary porewater". *Geol. Soc. of Amer. Bull.,* **86**:109-110.

HUNTER, R.J. (1989) *Foundation of colloid science.* Vol.1, Clarendon Press.

INGRAHAM, N.L. and SHADEL, C. (1992) "A comparison of the toluene distillation and vacuum/heat methods for extracting soil water for stable isotopic analysis". *J.Hydrol.,* **140**:371-387.

IYER, B. (1990) "Porewater extraction: comparison of saturation extract and high pressure squeezing". In: "Physico-chemical aspects of soil and related materials", ASTM Spec. Publ. 1095:159-170.

JACKSON, J.A. (Ed.) (1997) *Glossary of Geology.* American Geological Institute, Alexandria, Virginia, IV Ed., 769.

JAHNKE, R. A. (1988) "A simple, reliable, and inexpensive porewater sampler". *Limn. and Ocean.,* **33**:483-487.

JAMES, R.O. and PARKS, G.A. (1982) "Characterisation of aqueous colloids by their electric double layer and intrinsic surface chemical properties". *Surf. and Coll. Sci.,* **12**:119-216.

JONES, B. F., VANDENBURGH, A. S., TRUESDELL, A. H. and RETTIG, S. L. (1969) "Interstitial brines in playa sediments". *Chem. Geol.,* **4**:253-262.

JOSEPH, A.F. and MARTIN, F.J. (1923) "The moisture equivalent of heavy soils" *J. Agr. Sci.,* **13**:49-59.

JUSSERAND, C. (1979) "Une nouvelle technique pour l'étude isotopique de l'oxygène-18 des eaux présentes dans les sédiments et les sols". *C. R. Acad. Sci. Paris,* **228** série D:1027-1030.

JUSSERAND, C. (1980) " Extraction de l'eau interstitielle des sédiments et des sols. Comparaison des valeurs de l'oxygène-18 par différentes méthodes. Premiers résultats". *Catena,* 7:87-96.

KAHR, G., KRAEHENBUEHL, F., STOECKLI, H.F. and MULLER-VONMOOS, M. (1990) "Study of the water-bentonite system by vapor adsorption, immersion calorimetry and X-ray techniques: II. Heats of immersion, swelling pressures and thermodynamic properties". *Clay Min.,* **25**:499-506.

KALIL, E.K. and GOLDHABER, M. (1973) "A sediment squeezer for removal of porewater without air contact". *J. Sedim. Petrol.,* **43**/2:553-557.

KAPPES, T., SCHNIERLE, P. and HAUSER, P.C. (1997) "Potentiometric detection of inorganic anions and cations in capillary electrophoresis with coated-wire ion-selective electrodes". *Anal. Chim. Acta,* **350**:141-147.

KARABORNI, S., SMIT, B., HEIDUG, W., URAI, J. and VAN OORT, E. (1996) "The swelling of clays: molecular simulations of the hydration of montmorillonite". *Science,* **271**:1102-1104.

KARNLAND O. (1997) "Bentonite swelling pressure in strong NaCl solutions. Correlation between model calculations and experimentally determined data". SKB, TR 97-31:30.

KAUFMAN, R., LONG, A., BENTLEY, H.W. and DAVIS, S.N. (1984) "Natural chlorine isotope variations". *Nature,* **309**:338-340.

KENYON, B., KLEINBERG, R., STRALEY, C., GUBELIN, G. and MORRIS C. (1995) "Nuclear Magnetic Resonance imaging, Technology for the 21st century". *Oilfield Review,* 19-33.

KHARAKA, Y.K. and BERRY, F.A.F. (1973) "Simultaneous flow of water and solutes through geological membranes, I. Experimental investigation". *Geoch. Cosmoch. Acta,* **37**:2577-2603.

KHARAKA, Y.K. and SMALLEY, W.C. (1976) "Flow of water and solutes through compacted clays". *A.A.P.G. Bull.,* **60**/6:973-980.

KINNIBURGH, D.G., GALE, I.N., GOODDY, D.C., DARLING, W.G., MARKS, R.J., GIBBS, B.R., COLEBY, L.M., BIRD, M.J. and WEST, J.M. (1996) "Denitrification in the unsaturated zone". Brit. Geol. Surv. Tech. Rep. WD/96/63.

KINNIBURGH, D.G. and MILES, D.L. (1983) "Extraction and chemical analysis of interstitial water from soils and rocks". *Environ. Sci. Technol.,* **17**:362-368.

KITTRICK, J.A. (1980) "Gibbsite and kaolinite solubilities by immiscible displacement of equilibrium solutions". *Soil Sci. Soc. Am. J.,* **44**:139-142.

KITTRICK, J.A. (1983) "Accuracy of several immiscible displacement liquids". *Soil Sci. Soc. Am. J.,* **47**:1045-1047.

KOIKE, I. and TERAUCHI, K. (1996) "Fine scale distribution of nitrous oxide in marine sediments". *Mar. Chem.,* **52**:185-193.

KRAEHENBUEHL, F., STOECKLI, H.F., BRUMMER, F., KAHR, G. and MULLER-VONMOOS, M. (1987) "Study of the water-bentonite system by vapor adsorption, immersion calorimetry and x-ray techniques: I. Micropore volumes and internal surface areas, following Dubinin's theory". *Clay Min.,* **22**:1-10.

KRIUKOV, P.A. (1947) "Recent methods for physicochemical analysis of soils: Methods for separating soil solution". In: *Rukovodstvo dlya polevykh I laboratornykh issledovanii pochv,* Moscow Izdat, Akad, Nauk, SSSR, 3-15. (in Russian, cited in Kriukov and Manheim, 1982)

KRIUKOV, P.A. and MANHEIM, F.T. (1982) "Extraction and investigative techniques for study of interstitial waters of unconsolidated sediments: A review". In: *The Dynamic Environment of the Ocean Floor* (Fanning, K. A. and Manheim, F. T. Eds.), Heath,1-26.

KRIUKOV, P.A., ZHUCHKOVA, A.A. and RENGARTEN, E.V. (1962) "Change in the composition of solution preserved from clays and ion exchange". *Earth Science,* **144**:167-169.

LAMBE, T. W. and WHITMAN, R.V. (1979) *Soil Mechanics, SI Version.* John Wiley and Sons, New York.

LAWRENCE, J.R. and GIESKES, J.M. (1981) "Constraints on water transport and alteration in the oceanic crust from the isotopic composition of porewater". *J. Geophys. Res,* **86**/B9:7924-7934.

LEANEY, F.W., SMETTEM, K.R.J. and CHITTLEBOROUGH, D.J. (1993) "Estimating the contribution of preferential flow to subsurface runoff from a hill slope using deuterium and chloride". *J. of Hydrol.,* **147**:83-103.

LEENHEER, J.A. (1981) "Comprehensive approach to preparative isolation and fractionation of dissolved organic carbon from natural waters and waste waters". *Environ. Sci. and Techn.,* **15**:578-587.

LESSARD, G. and MITCHELL, J.K. (1985) "The causes and effect of ageing in quick clays". *Can. Geotech. J.,* **22**:335- 346.

LEWIS, R.J. Jr. (1993) *Hawley's Condensed Chemical Dictionary.* Van Norstrand Reinhold Company, New York, XII Ed., 1275.

LIPMAN, C. P. (1918) "A new method of extracting the soil solution". *University of California Publ. on Agricultural Science,* **3**:131-134.

LISITSYN, A.K., KIREEV, A.M., SOLODOV, I.N. and TOKAREV, N.I.(1984) "Using borehole cores for detailed investigation of the composition and properties of groundwater solutions". *Lith. and Min. Res.,* **19**/3:260-270.

LITAOR, M.I. (1988) "Review of soil solution samplers". *Wat. Res. Res.,* **24**/5:727-733.

LOW, P.F. (1982) "Water in clay-water systems". *Agronomie,* **2**/10:909-914.

LUSCZYNSKI, N. J. (1961) "Filter-press method of extracting water samples for chloride analysis". US Geol. Surv., Water Supply Paper 1544-A, 1-8.

LYONS, W.B., GAUDETTE, H.E. and SMITH, G.M. (1979) "Porewater sampling in anoxic carbonate sediments: oxidation artefacts". *Nature,* **277**:48-49.

MAGARA, K. (1974) "Compactation, ion filtration and osmosis in shale and their significance in primary migration". *A.A.P.G. Bull.,* **58**/2:283-290.

MAGARA, K. (1976) "Water expulsion from clastic sediments during compaction, directions and volumes". *A.A.P.G. Bull.,* **60**:543-553.

MANGELSDORF, P.C., WILSON, T.R.S. and DANIELL, E. (1969) "Potassium enrichments in interstitial waters of recent marine sediments". *Science,* **165**:171-174.

MANHEIM, F.T. (1966) "A hydraulic squeezer for obtaining interstitial water from consolidated and unconsolidated sediments". US Geol. Surv. Prof. Pap. 550-C:256.

MANHEIM, F.T. (1974) "Comparative studies on extraction of sediment interstitial waters: discussion and comment on the current state of interstitial water studies". *Clays and Clay Min.,* **22**:337-343.

MANHEIM, F.T. (1976) "Interstitial waters from marine sediments". In: *Chemical Oceanography* (Riley J.P. and Chester R. Eds.), Academic Press, New York, 115-186.

MANHEIM, F.T. and SAYLES, F.L. (1974) "Composition and origin of interstitial waters of marine sediments, based on Deep Sea Drilling cores". In: *The Sea* (Goldberg, E.D. Ed.) Wiley-Interscience, New York, 527-568.

MARSHALL, T.J. and HOLMES, J.W. (1979) *Soil physics.* Cambridge University Press, 374.

MARCUS, Y. (1985) *Ion solvation.* John Wiley and Sons, 306.

MATHIEU, R. and BARIAC, T. (1995) "Étude isotopique de l'évaporation de l'eau dans un sol argileux, expérimentation et modélisation". *Hydrogéologie,* **4**:85-97.

MAURICE, P., NAMJESNIK-DEJANOVIC, K., LOWER, S., PULLIN, M., CHIN, Y.P. and AIKEN, G.R. (1998) "Sorption and fractionation of natural organic matter on kaolinite and goethite". *Water Rock Interaction WRI 9 New Zealand, 30 March-3 April 1998* (Arehart G.B. and Huston J.R. Eds.), 109-113.

MAY, H.M., KINNIBURGH, D.G., HELMKE, P.A. and JACKSON, M.L. (1986) "Aqueous dissolution, solubilities and thermodynamic stabilities of common aluminosilicate clay minerals: Kaolinite and smectites". *Geoch. Cosmoch. Acta,* **50**:1667-1677.

MAYER, L.M. (1976) "Chemical water sampling in lakes and sediments with dialysis bags". *Limnol. Oceanogr.,* **21**:909-912.

McEWAN, D.M.C. and WILSON, M. (1980) "Interlayer and intercalation complexes of clay minerals". In: *Crystal Structures of Clay Minerals and their X-ray Identification.* Mineral Soc. Monogr., 5 (Brindley G.W. and Brown G. Eds.), 197-248.

McEWEN, T., CHAPMAN, N. and ROBINSON, P. (1990) "Review of data requirements for groundwater flow and solute transport modelling and the ability of site investigation methods to meet these requirements". DOE Rep., DOE/HIMP IG 2405-1.

METCALFE, R., ROSS, C.A.M., CAVE, M.R., GREEN, K.A., REEDER, S. and ENTWISLE, D.C. (1990) "Porewater and groundwater geochemistry at the Down Ampney fault research site". Brit. Geol. Surv. Tech. Rep, WE/90/46:52.

MILLOT, G. (1964) *Géologie des argiles; altérations, sédimentologie, géochimie.* Masson, Paris.

MOONEY, R.W., KENNAN, A.G. and WOOD, L.A. (1952a) "Adsorption of water vapor by montmorillonite. I. Heat of adsorption and application of the BET theory". *J. Amer. Chem. Soc.,* **74**:1367-1371.

MOONEY, R.W., KENNAN, A.G. and WOOD, L.A. (1952b) "Adsorption of water vapor by montmorillonite. II. Effect of exchangeable ions and lattice swelling as measured by X-ray diffraction". *J. Amer. Chem. Soc.,* **74**:1371-1374.

MOREAU-LE GOLVAN, Y. (1997) *Traçage isotopique naturel des transferts hydriques dans un milieu argileux de très faible porosité: les argilites de Tournemire (France).* Thesis, Université de Paris-Sud, 170.

MOREAU-LE GOLVAN, Y., MICHELOT, J.-L. and BOISSON, J.-Y. (1997) "Stable isotope contents of porewater in a claystone formation (Tournemire, France): assessment of the extraction technique and preliminary results". *Appl. Geoch.,* **12**/6:739-745.

MORGENSTERN, N.R. and BALASUBRAMONIAN, B.I. (1980) "Effects of pore fluid on the swelling of clay shale". In: *Proc. of the Fourth International Conference on Expansive Soils,* Denver, 1, 190-205.

MORTLAND, M.M. and RAMAN, K.V. (1968) "Surface acidity of smectites in relation to hydration, exchangeable cation and structure". *Clays and Clay Min.,* 16**:**393-398.

MOSS, D.P. and EDMUNDS, W.M. (1992) "Processes controlling acid attenuation in the unsaturated zone of a Triassic sandstone aquifer (U.K.) in the absence of carbonate minerals". *Appl. Geoch.,* **7**:573-583.

MOTELLIER, S., MICHELS, M.H., DUREAULT, B. and TOULHOAT, P. (1993) "Fiber optic pH sensors for in situ applications". *Sensors and actuators B,* **11**:467-473.

MOTELLIER, S., NOIRE, M.H., PITSCH, H. and DUREAULT, B. (1995) "pH determination of clay interstitial water using a fiber-optic sensor". *Sensors and actuators B,* **29**:105-115.

MUBARAK, A. and OLSEN, R.A. (1976) "Immiscible displacement of the soil solution by centrifugation". *Soil Sci. Soc. Am. J.,* **40**:329-331.

MUURINEN, A. and LEHIKOINEN, J. (1998) "Evolution of the porewater chemistry in compacted bentonite". In: *Scientific Basis for Nuclear Waste Management XXI, Mat. Res. Soc. Symp. Proc.,* **506**:415-422.

NAGENDER NATH, B., RAJARAMAN, V.S. and MUDHOLKAR, A.V. (1988) "Modified interstitial water squeezer for trace metal analysis". *Indian J. of Mar. Sci.,* **17**:71-72.

NAGRA (1997) "Geosynthese Wellenberg 1996: Ergebnisse der Untersuchungsphasen I und II". NAGRA, NTB 96-01.

NEA/SEDE (Working Group on measurement and Physical understanding of Groundwater flow through argillaceous media) (1998) "Catalogue of Characteristics of Argillaceous rocks". Int. Doc., Cd Rom publication.

NEILSON, G.W. and ENDERBY, J.E. (1989) "The coordination of metal aquaions". In: *Advances in Inorganic Chemistry,* 34 (Sykes A.G. Ed.), Academic Press, 195-218.

NEMECZ, E. (1981) *Clay Minerals.* Akademiai Kiado, Budapest Hungary.

NEUZIL, C.E. (1995) "Characterisation of flow properties, driving forces and porewater chemistry in the ultra-low permeability Pierre Shale, North America". *Proc. of an Int. Workshop on "Hydraulic and hydrochemical characterisation of argillaceous rocks"*, Nottingham, U.K., 7-9 June 1994, OECD/NEA, 65-74.

NORRISH, K. (1954) "The swelling of montmorillonite". *Trans. Faraday Soc.,* **18**:120-133.

NORTHRUP, Z. (1918) "The true soil solution". *Science,* **47**:638-639.

NOYNAERT, L., VOLCKAERT, G., DE CANNIÈRE, P., MEYNENDONCKX, P., LABAT, S., BEAUFAYS, R., PUT, M., AERTSENS, M., FONTEYNE, A. and VANDERVOORT, F. (1997) "The CERBERUS Project. A demonstration test to study the near field effects of a HLW-canister in an argillaceous formation". ONDRAF/NIRAS Nirond, 97-03.

NOVAKOWSKI, K.S. and VAN DER KAMP, G. (1996) "The radial diffusion method. 2. A semianalytical model for the determination of effective diffusion coefficients, porosity and adsorption". *Wat. Res. Res.,* **32**/6:1823-1830.

OHTAKI, H. and RADNAI, T. (1993) "Structure and dynamics of hydrated ions". *Chem. Rev.,* **93**/3:1157-1204.

OREM, W.H. and HATCHER, P.G. (1987) "Solid state ^{13}C NMR studies of dissolved organic matter in porewaters from different depositional environments". *Org. Geochem.,* **11**:73-82.

OSCARSON, D.W. and DIXON, D.A. (1989) "Elemental, mineralogical and pore-solution compositions of selected Canadian clays". AECL-9891:18.

OSENBRÜCK, K. (1996) *Age and dynamics of deep groundwater: a new method to analyse noble gases in the porewater of rock samples.* Thesis, University of Heidelberg (in German).

OSENBRÜCK, K. and SONNTAG, C. (1995) "A new method for the investigation of noble gases dissolved in porewater of sedimentary rocks. First results from coal and salt mines in Germany". In: *Isotope techniques in water resources management, Proc. of a Symp., Vienna, 20-24 March 1995,* IAEA, Vol.1, SM336/101P, 106-107.

OSENBRÜCK, K., LIPPMANN, J. and SONNTAG, C. (1998) "Dating very old porewaters in impermeable rocks by noble gas isotopes". *Geoch. Cosmoch. Acta,* **62**:3041-3045.

PARKER, F.W. (1921) "Methods of studying the concentration and composition of soil solution". *Soil Science,* **12**:209.

PARKHURST, D.L. (1995) "User's guide to PHREEQC: a computer program for speciation, reaction-path, advective transport, and inverse geochemical calculations". US Geol. Surv. Water Resources Investigations, 95-4227:143.

PARKHURST, D.L., THORSTENSON, D.C. and PLUMMER, L.N. (1980) "PHREEQE: a computer program for geochemical calculations". US Geol. Surv. Water Resources Investigations, 80-96, 210.

PARSHIVA MURTHY, A.S. and FERREL, R.E. Jr. (1972) "Comparative chemical composition of sediment interstitial waters". *Clays and Clay Min.,* **20**:317-321.

PARSHIVA MURTHY, A.S. and FERREL, R.E. Jr. (1973) "Distribution of major cations in estuarine sediments". *Clays and Clay Min.,* **21**:161-165.

PATTERSON, R. J., FRAPE, S. K., DYKES, L. S. and McLEOD, R. A. (1978) "A coring and squeezing technique for the detailed study of subsurface chemistry". *Can. J. of Earth Sci.,* **15**:162-169.

PEARSON, F.J. (1998) "Geochemical and other porosity types in clay-rich rocks". *Water Rock Interaction WRI 9, New Zealand, 30 March-3 April 1998* (Arehart G.B. and Huston J.R. Eds.), 259-262.

PEARSON, F.J. (in press) "What is the porosity of a mudrock?". In: *Muds and mudstones: Physical and fluid flow properties.* Geological Society, Special Publication, (Aplin A.C., Fleet A.F. and Macquaker J. Eds.), London, 158, 9-21.

PEARSON, F.J. and BERNER, U. (1991) "Nagra thermochemical database I. Core data". NAGRA, NTB 91-17.

PEARSON, F.J., BERNER, U. and HUMMEL, W. (1992) "Nagra thermochemical database II". Supplemental data 05/92. NAGRA, NTB 91-18.

PEARSON, F.J., *et al.* (in preparation) "Hydrochemistry". In: *Mont Terri Rock Laboratory: Results of the hydrogeological, geochemical and geotechnical experiments performed in 1996 and 1997. Landeshydrologie und -geologie, Geologische Berichte,* (Thury, M. and Bossart, P. Eds.), 23.

PEARSON, F.J., FISHER, D.W. and PLUMMER, L.N. (1978) "Correction of ground-water chemistry and carbon, isotopic composition for effects of CO_2 outgassing". *Geoch. Cosmoch. Acta,* **42**:1799-1807.

PETERS, C.A., YANG, I.C., HIGGINS, J.D. and BURGER, P.A. (1992) "A preliminary study of the chemistry of porewater extracted from tuff by one dimensional compression". *Water-Rock Interaction WRI-7* (Kharaka, Y.K. and Maest, A., Eds.), **1**:741-745.

PHILLIPS, F.M. and BENTLEY, H.W. (1987) "Isotopic fractionation during ion filtration: I. Theory". *Geoch. Cosmoch. Acta,* **51**:683-695.

PITSCH, H., L'HENORET, P., BOURSAT, C., DE CANNIÈRE, P. and FONTEYNE, A. (1995a) "Characterization of deep underground fluids, Part II: Redox potential measurements". *Proc. of the 8th Int. Symp. on Water-Rock Interaction WRI 8,* Vladivostok, Russia, 15-19 August 1995 (Kharaka, Y.K. and Chudaev, O.V. Eds.), Balkema, Rotterdam, 471-475.

PITSCH, H., MOTELLIER, S., L'HENORET, P. and BOURSAT, C. (1995b) "Characterization of deep underground fluids, Part I: pH determination in a clayey formation". *Proc. of the 8th Int. Symp. on Water-Rock Interaction WRI-8,* Vladivostok, Russia, 15-19 August 1995 (Kharaka Y.K. and Chudaev O.V. Eds.), Balkema, Rotterdam, 467-470.

PITSCH, H., STAMMOSE, D., KABARE, I. and LEFEVRE, I. (1992) "Sorption of major cations on pure and composite clayey materials". *Appl. Clay Sci.,* **7**:239-243.

PLUMMER, L.N. (1992) "Geochemical modeling of water-rock interaction: past, present, future". *Water-Rock Interaction WRI-7* (Kharaka, Y.K. and Maest, A., Eds.), 1, Balkema, Rotterdam, 23-33.

PLUMMER, L.N., PRESTEMON, E.C. and PARKHURST, D.L. (1992) "NETPATH: an interactive code for interpreting NET geochemical reactions for chemical and isotopic data along a flow PATH". *Water-Rock Interaction WRI-7,* (Kharaka, Y.K. and Maest, A., Eds.), Balkema, Rotterdam, 239-242.

POUTOUKIS, D. (1991) *Hydrochimie, teneurs isotopiques et origine des saumures associées aux gisements de potasse d'Alsace.* Thesis, Université de Paris-Sud, 140.

PRESLEY, B.J., BROOKS, R.R. and KAPPEL, H.M. (1967) "A simple squeezer for removal of interstitial water from ocean sediments". *J. Mar. Res.,* **25**:355-357.

PROST, R. (1975) *Étude de l'hydratation des argiles: interaction eau-minéral et mécanisme de la retention de l'eau.* Thesis, Université Pierre et Marie Curie, Paris, 535.

PUSCH, R. and KARNLAND, O. (1986) "Aspects of the physical state of smectite adsorbed water". SKB Tech. Rep., 86-25.

PUSCH, R., KARNLAND, O. and HOKMARK, H. (1990) "GMM: A general microstructural model for qualitative and quantitative studies of smectite clays". SKB Tech. Rep., 90-43.

PUSCH, R., KARNLAND, O. and MUURINEN, A. (1989) "Transport and microstructural phenomena in bentonite clay with respect to the behaviour and influence of Na, Cu and U". SKB Tech. Rep., 89-34.

QUIRK, J.P. and AYLMORE, L.A.G. (1971) "Domains and quasi-crystalline regions in clay systems". *Soil Sci. Soc. Am. J.,* 35:652-654.

RAMANN, E., MÄRZ, S. and BAUER, H. (1916) "Uber boden-press-säfte". *Int. Mitt. für Bodenkunde,* **6**:1:27-34.

REEBURGH, W. S. (1967) "An improved interstitial water sampler". *Limn. and Ocean.,* **12**:163-165.

REEDER, S., CAVE, M.R., BATH, A.H., ENTWISLE, D.C., INGLETHORPE, S.J., PEARCE, J.M., TRICK, J.K., BLACKWELL, P.A. and GREEN, K.A. (1993) "A study of the Boom clay drillcore from Mol in Belgium. Chemical and isotopic characterisation of porewater and clay mineralogy". Brit. Geol. Surv. Tech. Rep. WI/93/12C

REEDER, S., CAVE, M.R., ENTWISLE, D.C., TRICK, J.K., HARMON, K.A., BLACKWELL, P.A., MITCHELL, N. and COOK, J.M. (1992) "The extraction and analysis of pore fluids from the Boom clay drillcore, Mol, Belgium". Brit. Geol. Surv. Tech. Rep. WI/92/7C

REEDER, S., CAVE, M.R., ENTWISLE, D.C. and TRICK, J.K. (1998) "Extraction of water and solutes from clayey material: a review and critical discussion of available techniques". Brit. Geol. Surv. Tech. Rep., WI/98/4C, 60.

REEDER, S., CAVE, M.R., ENTWISLE, D.C., TRICK, J.K., BLACKWELL, P.A., WRAGG, J. and BURDEN, S. (1997) "Clay porewater characterisation: New investigations at the Gard MAR 402 site". ANDRA Tech. Rep. D RP O BGS 97-002.

REVESZ, K. and WOODS, P.H. (1990) "A method to extract soil water for stable isotope analysis". *Journ. Hydrol.,* **115**:397-406.

RHOADES, J.D. (1982) "Cation exchange capacity". In: *Methods for soil analysis (Agronomy)* **9**:149-157.

RICARD, P. (1993) *Étude isotopique des fluides interstitiels et des minéraux de fracture dans les argilites toarciennes de Tournemire (Aveyron).* Thesis, Université de Paris-Sud.

RICHARDS, L.A. (1941) "A pressure-membrane extraction apparatus for soil solutions". *Soil Sci.,* **51**:377-386.

RICHARDS, L.A. (Ed.) (1954) "Diagnosis and improvement of saline and alkali soils". U.S.D.A. Agriculture Handbook 60, US Government Printing Office, Washington.

ROBINSON, R.A. and STOKES, R.H. (1970) *Electrolyte solutions,* 2nd ed. Academic Press, New York, 559.

RODVANG, S.J. (1987) *Geochemistry of the weathered zone of a fractured clayey deposit in Southwestern Ontario.* M.Sc. thesis, University of Waterloo, Ontario, 177.

ROSENBAUM, M. (1976) "Effect of compactation on the pore fluid chemistry of montmorillonite". *Clays and Clay Min.,* **24**:118-121.

ROSS, D.S. and BARTLETT, R.J. (1990) "Effects of extraction methods and sample storage on properties of solutions obtained from forested spodsols". *J. Environ. Qual.,* **19**:108-113.

RÜBEL, A., ROGGE, T., LIPPMANN, J. and SONNTAG, C. (1998) "Two methods to extract porewater of clay stones for stable isotope analysis and the determination of water content: vacuum distillation and direct equilibration". Abs. Atelier thématique 98/1, GdR ForPro, Grenoble, France, 27/02/98.

SAUZAY, G. (1974) "Sampling of lysimeters for environmental isotopes of water". *Proc. Symp. Isotope Techniques in Groundwater Hydrology,* IAEA, Vienna, 11-15/03/1974, II, 61-68.

SAWHNEY, B.L. (1972) "Selective sorption and fixation of cations by clay minerals: a review". *Clays and Clay Min.,* **20**:93-100.

SAXENA, R.K. and DRESSIE, Z. (1983) "Estimation of groundwater recharge and moisture movement in sandy formations by tracing natural oxygen-18 and injected tritium profiles in the unsaturated zone". *Proc. of a Meeting "Isotope Hydrology 1983", IAEA, Vienna, 12-16 September 1983.*

SAYLES, F.L., MANGELSDORF, P.C., WILSON, T.R.S. and HUME, D.N. (1976) "A sampler for the in situ collection of marine sedimentary porewater". *Deep-sea Res.,* **23**:259-264.

SCHAFFER, R.J., WALLACE, J. and GARWOOD, F. (1937) "The centrifuge method of investigating the variation of hydrostatic pressures with water content in porous materials". *Trans. Faraday Soc.,* **33**:723-734.

SCHLINDER, P.W. and GAMSJAGER, H. (1972) "Acid base reactions of the TiO_2 (Anatase) water interface and the point of zero charge of TiO_2 suspensions". *Kolloid Z. Z. Polymere,* **250**:759-763.

SCHLINDER, P.W. and KAMBER, H.R. (1968) "Die acidität von silanolgruppen". *Helv. Chim. Acta,* **51**:1781-1786.

SCHLUMBERGER (1997) *Italy 2000.* CD Rom publication, 124.

SCHMIDT, G.W. (1973) "Interstitial water composition and geochemistry of deep Gulf Coast shales and sandstones". *A.A.P.G. Bull.,* **57**/2:321-337.

SCHNITZER, M. (1978) "Humic substances: chemistry and reactions". In: *Soil organic matter, Developments in soil science,* 8, Elsevier, 1-64.

SHATKAY, M. and MARGARITZ, M. (1987) "Dolomitization and sulphate reduction in the mining zone between brine and meteoric water in the newly exposed shores of the Dead Sea". *Geoch. Cosmoch. Acta,* **51**:1135-1141.

SHEPPARD, S.M.F. and GILG, H.A. (1996) "Stable isotope geochemistry of clay minerals". *Clay Minerals,* **31**:1-24.

SHOLKOVITZ, E. (1973) "Interstitial water chemistry of the Santa Barbara Basin sediments". *Geoch. Cosmoch. Acta,* **37**:2043-2073.

SIEVER, R. (1962) "A squeezer for extracting interstitial water". *J. Sedim. Petrol.,* **32**:329-331.

SKEMPTON, A.W. (1944) "Notes on the compressibility of clays". *Quart. J. of the Geol. Society,* **100**:119-135.

SMITH, D.B., WEARN, P.L., RICHARDS, H.J. and ROWE, P.C. (1970) "Water movement in the unsaturated zone of high and low permeability strata by measuring natural tritium". *Proc. of the Symp. "Isotope Hydrology 1970", IAEA, Vienna*, 73-87.

SOFER, Z. (1978) "Isotopic composition of hydration water in gypsum". *Geoch. Cosmoch. Acta,* **42**:1141-1149.

SOFER, Z. and GAT, J.R. (1975) "The isotope composition of evaporating brines: effect of the isotopic activity ratio in saline solutions". *Earth Plant. Sci. Lett,* **26**:179-186.

SOFER, Z. and GAT, J.R. (1972) "Activities and concentrations of Oxygen-18 in concentrated aqueous salt solutions: analytical and geophysical implications". *Earth Plant. Sci. Lett,* **15**:232-238.

SPOSITO, G. (1980) "Derivation of the Freundlich equation for ion exchange reactions in soils". *Soil Sci. Soc. Am. J.,* **44**:652.

SPOSITO, G. (1984) *The surface chemistry of soils.* Oxford University Press, New York, 234.

SPOSITO, G. (1989) *The chemistry of soils.* Oxford University Press, 277.

SPOSITO, G. (1992) "The diffuse-ion swarm near smectite particles suspended in 1:1 electrolyte solutions: modified Gouy-Chapman theory and quasicrystal formation". In: *Clay water interface and its rheological implications,* CMS Workshop Lecures Vol. 4 (Güven N. and Pollastro R.M. Eds.), The Clay Minerals Society, 127-156.

SPOSITO, G. and PROST, R. (1982) "Structure of water adsorbed on smectites". *Chem. Reviews,* **82**/6:553-573.

STEINMANN, P. and SHOTYK, W. (1996) "Sampling anoxic porewaters in peatlands using "peeper" for in situ filtration". *Fresenius J. Anal. Chem.,* **354**:708-713.

STERN, O. (1924) "Zur Theorie der Elektrolytischen Doppleschriht". *Zietsch. Elektrochem.,* **30**:508-516.

STEWART, G.L. (1972) "Clay-water interaction, the behaviour of ^3H and ^2H in adsorbed water and the isotope effect". *Soil Sci. Soc. Am. Proc.,* **36**:421-426.

STROBEL, J., WIN, P., WOUTERS, L. and HAGOOD, M. (1998) "Groundwater permeability correlations in the Boom clay using wireline logs". In: *Proc. of the Int. High-Level Nuclear Waste Conference,* Las Vegas (in press).

STUDER, R. (1961) "Méthode de détermination des réserves hydriques des sols". *Ann. Agron.,* **12**/6:599- 608.

STUMM, W., HUANG, C.P. and JENKINS, S.R. (1970) "Specific chemical interactions affecting the stability of dispersed systems". *Croat. Chem. Acta,* **42**:223-244.

STUMM, W. and MORGAN, J.J. (1996) *Aquatic chemistry. Chemical equilibria and rates in natural waters.* Wiley Interscience, III Ed., 1022.

SUN, Y., LIN, H. and LOW, P.F. (1986) "The non specific interaction of water with the surfaces of clay minerals". *J. Coll. Interf. Sci.,* **112**/2:556-564.

SUQUET, H., DE LA CALLE, C. and PEZERAT, H. (1975) "Swelling and structural organization of saponite". *Clays and Clay Min.,* **23**:1-9.

SUQUET, H., IIYAMA, J.T., KODAM, H. and PEZERAT, H. (1977) "Synthesis and swelling properties of saponites with increasing layer charge". *Clays and Clay Min.,* **25**:231-242.

SUQUET, H. and PEZERAT, H. (1987) Parameters influencing layer stacking types in saponite and vermiculite: a review". *Clays and Clay Min.*, **35**:353-362.

SWALAN, J.J. and MURRAY, J.W. (1983) "Trace metal remobilization in the interstitial water of red clay and hemipelagic marine sediments". *Earth and Planet. Sci. Lett.*, **64**:213-230.

SWARTZEN-ALLEN, S.L. and MATIJEVIC, E. (1974) "Surface and colloid chemistry of clays". *Chem. Rev.*, **74**:385-400.

TARDY, Y. and TOURET, O. (1987) "Hydration energies of smectites: a model for glauconite, illite, and corrensite formation". In: *Proc. of the Int. Clay Conf. 1985 (*Schultz L.G., Van Olphen H. and Mumpton F.A. Eds.), Clay Minerals Soc., Bloomington, Indiana, 46-52.

TAUBE, H. (1954) "Use of oxygen isotope effects in the study of hydration of ions". *J. Phys. Chem.*, **58**:523-528.

TEASDALE, P.R., BATLEY, G.E., APTE, S. and WEBSTER, I.T. (1995) "Porewater sampling with sediment peepers". *Trends in Analytical Chem.*, **14**/6:250-256.

TESSIER, D. and PEDRO, G. (1987) "Mineralogical characterisation of 2:1 clays in soils: Importance of the clay structure". In: *Proc. of the Int. Clay Conf. 1985 (*Schultz L.G., Van Olphen H. and Mumpton F.A. Eds.), Clay Minerals Soc., Bloomington, Indiana, 78-84.

TESTINI, C. and GESSA, C. (1989) "Le fasi solide". In: *Chimica del suolo*, Patron, Bologna, 627.

THENG, B.K.G. (1974) *The chemistry of Clay-Organic reactions.* John Wiley and Sons, 343.

THOMA, G., ESSER, N., SONNTAG, C., WEISS, W., RUDOLPH, J. and LÉVÊQUE, P. (1978) "New technique of in-situ soil-moisture sampling for environmental analysis applied at Pilat sand dune near Bordeaux: HETP modelling of bomb tritium propagation in the unsaturated sand zones". *Isotope Hydrology 1978,* 2, IAEA, Vienna, 753-768.

THOMAS, G.W. (1977) "Historical developments in soils chemistry: ion exchange". *Soil Sci. Soc. Am. J.*, **41**:230 -238.

THOMAS, M.D. (1921) "Aqueous vapour pressure of soils" *Soil Science,* **11**:409

THURMAN, E.M. (1985a) *Organic geochemistry of natural waters.* Martinus Nijhoff/Dr. W. Junk Publ., The Hague.

THURMAN, E.M. (1985b) "Humic substances in groundwater". In: *Humic substances, I. Geochemistry, characterisation and isolation,* John Wiley, New York, 87-103.

THURY, M. and BOSSART, P. (Eds.) "Mont Terri Rock Laboratory: Results of the hydrogeological, geochemical and geotechnical experiments performed in 1996 and 1997". Landeshydrologie und -geologie, *Geologische Berichte,* 23 (in preperation).

TISSOT, B. and WELTE, D.H. (1978) *Petroleum formation and occurrence.* Springer Verlag.

TORSTENSSON, B.A. (1984) "A new system for groundwater monitoring". *Ground Water Mon. Rev.*, **4**/4:131-138.

TORSTENSSON, B.A. and PETSONK, A.M. (1985) "A hermetically isolated sampling method for groundwater investigations". *ASTM Symp. of Field Methods for Ground Water Contamination Studies and their Standardisation, Cocoa Beach, Florida, 2-7 Feb. 1985,* 35.

TOURET, O., PONS, C.H., TESSIER, D. and TARDY, Y. (1990) "Étude de la répartition de l'eau dans des argiles saturées Mg^{2+} aux fortes teneurs en eau". *Clay Minerals,* **25**:217-233.

TRAVIS, C.C. and ETNIER, E.L. (1981) "A survey of sorption relationships for reactive solutes in soils". *J. Environ. Qual.,* **10**:8.

TSHIBANGU, J.-P., SARDA, J.-P. and, AUDIBERT-HAYET, A. (1996) "Étude des interactions mécaniques et physicochimiques entre les argiles et les fluides de forage. Application à l'argile de Boom (Belgique)". *Rev. Inst. Franç. Pétrole,* **51**/4:497-526.

TURNER, J.V. and GAILITIS, V. (1988) "Single step method for hydrogen isotope ratio measurement of water in porous media". *Anal. Chem.,* **60**:1244-1246.

URBAN, N.R., DINKEL, C. and WEHRLI, B. (1997) "Solute transfer across the sediment surface of a eutrophic lake: I. Porewater profiles from dialysis samplers". *Aquatic Sci.,* **59**:1-25.

VAN DER KAMP, G., STEMPVOORT, D.R., WASSENAAR, L.I. (1996) "The radial diffusion method. 1. Using intact cores to determine isotopic composition, chemistry, and effective porosities for groundwater in aquitards". *Wat. Res. Res.,* **32**/6:1815-1822.

VAN OLPHEN, H. (1963) *An introduction to clay colloid chemistry for clay technologists, geologists and soil scientists.* Interscience Publishers, 301.

VAN OLPHEN, H. (1965) "Thermodyamics of interlayer adsorption of water in clays. I. Sodium vermiculite". *J. Collois Sci.,* **20**:822-837.

VAN OLPHEN, H. (1969) "Thermodyamics of interlayer adsorption of water in clays. II. Magnesian vermiculite". In: *Proc. of the Int. Clay Conference,* Tokyo. Israel University Press, 1, 649-657.

VAN OLPHEN, H. and FRIPIAT, J.J. (Eds.) (1979) *Data handbook for clay materials and other non-metallic minerals, providing those involved in clay research ans industrial application with sets of authoritative data describing the physical and chemical properties and mineralogical composition of the available reference materials.* Pergamon Press, 346.

VEIHMEYER, F.J., ISRAELSON, O.W. and CONRAD, J.P. (1924) Agr. Exper. Station, Berkeley California, Tech. Paper 16 (cited in Schaffer *et al.,* 1937)

VEITH, J.A. and SPOSITO, G. (1977) "The use of the Langmuir equation in the interpretation of "adsorption" phenomena". *Soil Sci. Soc. Am. J.,* **41**:679.

VELDE, B. (1992) *Introduction to clay minerals. Chemistry, origins, uses and environmental significance.* Chapman and Hall, 198.

VILLAR, M.V., LLORET, A., CUEVAS, J., YLLERA, A., FERNÁNDEZ, A.M. and JIMINEZ DE CISNEROS, C. (1997) "FEBEX project: laboratory experiments. In situ testing in underground research laboratories for radioactive waste". *Proc. of a cluster seminar, Alden Biesen, Belgium, 10-11 December 1997,* DOC XII/015/98, EN, 247-261.

VON DAMM, K.L. and JOHNSON, K.O. (1988) "Geochemistry of shale groundwaters: results of preliminary leaching experiments", ORNL/TM 10535, NTIS.

VUATAZ, F.D. (1987) " Diagraphie et suivis géochimiques des eaux souterraines: exemple de sondages dans le socle cristallin ". *Géothermie Actualités,* **4**/2:25-33.

VUATAZ, F.D., BRACH, M., CRIAUD, A. and FOUILLAC, C. (1990) "Geochemical monitoring of drilling fluids: a powerful tool to forecast and detect formation waters". Soc. Pet. Eng., Formation Evaluation June, 177-184.

WABER, H.N. and MAZUREK, M. (1998) "Tournemire shale porewater composition: derivation from geochemical modeling". University of Bern, Switzerland, Int. Rep., 29.

WADA, K. and OKAMURA, Y. (1977) "Measurements of exchange capacities and hydrolysis as a means of characterizing cation and anion retention by soils." In: *Proc. Intern. Seminar on Soil Environ. and Fertility Management in Intensive Agriculture, Society of Scientific Soil Manure, Tokyo, Japan,* 811-815.

WALKER, G.R., WOODS, P.H. and ALLISON, G.B. (1994) "Interlaboratory comparison of methods to determine the stable isotope composition of soil water". *Chem. Geol.,* **111**:297-306.

WALTERS, L.J. Jr. (1967) *Bound halogens in sediments.* Doctoral Thesis, Massachusetts Institute of Technology (cited in Kriukov and Manheim, 1982).

WEAVER, C.E. and BECK, K.C. (1971) "Clay water diagenesis during burial: how mud becomes gneiss". G.S.A. Spec. Publ., **134**:96.

WEISS, C.A. Jr. and GERASIMOWICZ, W.V. (1996) "Interaction of water with clay minerals as studied by ^{2}H nuclear magnetic resonance spectroscopy". *Geoch. Cosmoch. Acta,* **60**:265-275.

WESTALL, J., ZACHARY, J.L. and MOREL, F.M.M. (1976) "MINEQL, A computer code for the calculation of chemical equilibrium composition of aqueous systems". MIT Techn. Note No. 18, Boston, USA.

WHELAN, B.R. and BARROW, N.J. (1980) "A study of a method for displacing soil solution by centrifuging with an immiscible liquid". *J. Environ. Qual.,* **9**/2:315-319.

WIJLAND, G.C., LANGEMEIJER, H.D., STAPPER, R.A.M., MICHELOT, J.-L. and GLASBERGEN, P. (1991) "Evaluation and development of hydrological and geochemical investigation methods for aquifers of low permeability". CEC EUR 13125 EN, 96.

WILKINSON, M., FALLICK, A.E., KEANEY, G.M.J., HASZELDINER, S. and McHARDY, W.J. (1994) "Stable isotopes in illite: the case for meteoric water flushing within the upper Jurassic Fulmar Formation sandstones, U.K. North Sea". *Clay Minerals,* **29**:567-574.

WILSON, M.J. (Ed.) (1987) *A handbook of determinative methods in clay mineralogy.* Blackie, 308.

WIN P., STROBEL J., WOUTERS, L. and HAGOOD, M. (1998) "Definition of clay-silt sequences using wireline logs". In: *Proc. of the Int. High-Level Nuclear Waste Conference,* Las Vegas (in press).

WOLERY, T.J., JACKSON, K.J., BOURCIER, W.L., BRUTON, C.J., VIANI, B.E., KNAUSS, K.G. and DELANY, J.M. (1990) "Current status of the EQ3/6 software package for geochemical modeling". In: *Chemical modeling in aqueous systems II (*Melchior D.C. and Basset R.L. Eds.), *Am. Chem. Soc. Symp. Ser.,* **416**:104-116.

YANG, I.C., DAVIS, G.S. and SAYRE, T.M. (1990) "Comparison of porewater extraction by triaxial compression and high speed centrifugation methods". *Proc. Conf. on Minimizing Risk to the Hydrologic Environment, American Institute of Hydrology,* 250-259.

YANG, I.C., HIGGINS, J.D. and HAYES, L.R. (1995) "Porewater extraction from unsaturated tuff using triaxial and one-dimensional compression methods". *Proc. of an Int. Workshop on "Hydraulic and hydrochemical characterisation of argillaceous rocks", Nottingham, U.K., 7-9 June 1994,* OECD/NEA, 137-155.

YARIV, S. and CROSS, H. (1979) *Geochemistry of colloid systems for earth scientists.* Springer-Verlag, Berlin, 450.

YECHIELI, Y., MAGARITZ, M., SHATKAY, M., RONEN, D. and CARMI, I. (1993) "Processes affecting interstitial water in the unsaturated zone at the newly exposed shore of the Dead Sea, Israel". *Chem. Geol. (Isotope Geosc.),* **103**:207-225.

YEH, H.W. (1980) "D/H ratios and late stage dehydration of shales during burial". *Geoch. Cosmoch. Acta,* **44**:341-352.

YONG, R.N., MOHAMED, A.M.O., and WARKENTIN, B.P. (1992) "Principles of contaminant transport in soils". *Developments in Geotechnical Engineering,* Elsevier, **73**:327.

YONG, R. N. and WARKENTIN, B.P. (1975) "Soil properties and behaviour". In: *Developments in Geotechnical Engineering, 5,* Elsevier, Amsterdam.

YORMAH, T.B.R. and HAYES, M.H.B. (1993) "Sorption of water vapour by the Na^+-exchanged clay-sized fractions of some tropical soil samples". International Center for Theoretical Physics, Trieste (Italy) IC/93/334:15.

YOU, C.F., SPIVACK, A.J., GISKES, J.M., MARTIN, J.B. and DAVISSON, M.L. (1996) "Boron contents and isotopic compositions in porewaters: a new approach to determine temperature induced artifacts. Geochemical implications". *Marine Geol.,* **129**:351-361.

ZABOWSKI, D. and SLETTEN, R.S. (1991) "Carbon dioxide degassing effects on the pH of spodosol soil solutions". *Soil Sci. Soc. Am. J.,* **55**:1456-1461.

ZHENG, Y.-F. (1993) "Calculation of oxygen isotope fractionation in hydroxyl-bearing silicates". *Earth and Planet. Sci. Lett.,* **120**:247-263.

ZIMMERMANN, U., EHHALT, D. and MÜNNICH, K.O. (1966) "Soil-water movement and evapotranspiration: changes in the isotopic composition of the water". *Proc. of the Symp. "Isotopes in Hydrology",* IAEA, Vienna, 567-585.

ZIMMERMANN, U., MÜNNICH, K.O. and ROETHER, W. (1967) "Downward movement of soil moisture traced by means of hydrogen isotopes". *Geophys. Monogr.,* **11**:28-36.

OECD PUBLICATIONS, 2, rue André-Pascal, 75775 PARIS CEDEX 16
PRINTED IN FRANCE
(66 2000 02 1 P) ISBN 92-64-17181-9 – No. 51451 2000